NUREG-1650, Rev. 2

The United States of America Fourth National Report for the Convention on Nuclear Safety

Manuscript Completed: September 2007
Date Published: September 2007

September 2007

U.S. Nuclear Regulatory Commission
Washington, DC 20555-0001

THE UNITED STATES OF AMERICA

FOURTH NATIONAL REPORT

FOR THE

CONVENTION ON NUCLEAR SAFETY

SEPTEMBER 2007

U.S. NUCLEAR REGULATORY COMMISSION

WASHINGTON, DC 20555-0001

ABSTRACT

The United States Nuclear Regulatory Commission has updated NUREG-1650, Rev. 1, "The United States Third National Report for the Convention on Nuclear Safety," issued September 2004, and will submit this report for peer review at the fourth review meeting of the Convention on Nuclear Safety at the International Atomic Energy Agency in April 2008. This report addresses the safety of land-based commercial nuclear power plants in the U.S. It demonstrates how the U.S. Government achieves and maintains a high level of nuclear safety worldwide by enhancing national measures and international cooperation, and by meeting the obligations of all the articles established by the Convention. These articles address the safety of existing nuclear installations, the legislative and regulatory framework, the regulatory body, responsibility of the licensee, priority to safety, financial and human resources, human factors, quality assurance, assessment and verification of safety, radiation protection, emergency preparedness, siting, design and construction, and operation. New to this update is a contribution by the Institute of Nuclear Power Operations (INPO) describing work done by the U.S. nuclear industry to ensure safety. Since the prime responsibility for the safety of a nuclear installation rests with the license holder, it was decided to involve the nuclear industry collectively through INPO, in explaining how the nuclear industry maintains and improves nuclear safety.

CONTENTS

CONVENTION ON NUCLEAR SAFETY REPORT: THE ROLE OF THE INSTITUTE OF NUCLEAR POWER OPERATIONS IN SUPPORTING THE UNITED STATES COMMERCIAL NUCLEAR ELECTRIC UTILITY INDUSTRY'S FOCUS ON NUCLEAR SAFETY

EXECUTIVE SUMMARY

This is the "United States Fourth National Report for the Convention on Nuclear Safety" (NUREG-1650, Rev. 2). The third report was issued by the United States (U.S.) Nuclear Regulatory Commission (NRC) in September 2004. This report will be submitted for peer review at the fourth review meeting of the Convention on Nuclear Safety, to be convened at the International Atomic Energy Agency in April 2008. This report addresses the safety of land-based commercial nuclear power plants in the U.S. It demonstrates how the U.S. Government achieves and maintains a high level of nuclear safety worldwide by enhancing national measures and international cooperation, and by meeting the obligations of all the articles established by the Convention. These articles address the safety of existing nuclear installations, the legislative and regulatory framework, the regulatory body, responsibility of the licensee, priority to safety, financial and human resources, human factors, quality assurance, assessment and verification of safety, radiation protection, emergency preparedness, siting, design and construction, and operation.

This report addresses the issues identified in the peer review of the third review meeting in April 2005 and discusses challenges and issues that have arisen since that time. The NRC's challenges identified in the third review meeting were:

(1)   ensuring that the post-September 11, 2001, emphasis on security does not have an adverse effect on operational safety,

(2)   continuing the focus on safe long-term operation and emerging technical issues,

(3)   preparing for the potential licensing of new reactors,

(4)   appropriately emphasizing safety culture, safety management, and organizational aspects of nuclear safety,

(5)   continued monitoring and analysis of operating experience and timely implementation of lessons learned, and

(6)   expanding risk-informed, performance-based concepts while ensuring a consistent regulatory approach.

At the third review meeting, highlighted planned initiatives included:

(1)   enhancing the Reactor Oversight Process to increase NRC inspector review of licensee safety-conscious work environment,

(2)   working with the international community to define the ways and means to detect deteriorating performance, including safety culture and safety management, and

(3)   planning for the performance of an International Regulatory Review Team (renamed the Integrated Regulatory Review Service) self-assessment.

This report updates the status of these initiatives.

The safety issues discussed in the third U.S. National Report were reactor materials problems, pressurized-water reactor containment sump performance, electric grid reliability, emergency preparedness, and security. The status of these safety issues and those that have arisen since 2004 is discussed in this report, namely, reactor materials degradation problems, and pressurized-water reactor containment sump blockage due to post-loss-of-coolant accident chemical formation, and issues regarding post-power-uprate equipment.

Since new reactors are of increasing international interest, this report also discusses the NRC's preparations for licensing new reactors in the United States, an activity the NRC has not undertaken in the past 30 years.

The NRC considers managing human capital a major focus area, and this report discusses how the agency recruits and trains new staff, captures and passes on lessons learned from past events, and attempts to capture and pass on the knowledge of long-term staff as they prepare to retire.

For the first time, the Institute of Nuclear Power Operations (INPO) is providing input to the U.S. National Report. Since the prime responsibility for the safety of a nuclear installation rests with the license holder, it was decided to involve the nuclear industry collectively through INPO, in explaining how the nuclear industry maintains and improves nuclear safety.

PART 1

INTRODUCTION

This section describes the purpose and structure of the "United States Fourth National Report for the Convention on Nuclear Safety," the United States (U.S.) national policy toward nuclear activities, the main national nuclear programs, and the current nuclear safety issues. It also highlights major regulatory accomplishments since the previous (third) U.S. National Report was submitted in 2004 (see NUREG-1650, Rev. 1, "The United States Third National Report for the Convention on Nuclear Safety," September 2004).

Purpose and Structure of this Report

The United States of America is submitting this updated report for peer review to the Fourth Review Meeting of the Contracting Parties to the Convention on Nuclear Safety (the Convention). The scope of this report considers only the safety of land-based commercial nuclear power plants, consistent with the definition of nuclear installations provided in Article 2 and the scope of Article 3 of the Convention.

This report demonstrates how the U.S. Government meets the following objectives described in Article 1 of the Convention:

(i)   to achieve and maintain a high level of nuclear safety worldwide through the enhancement of national measures and international cooperation including, where appropriate, safety-related technical cooperation

(ii)   to establish and maintain effective defenses in nuclear installations against potential radiological hazards in order to protect individuals, society, and the environment from harmful effects of ionizing radiation from such installations

(iii)   to prevent accidents with radiological consequences and to mitigate such consequences should they occur

Technical and regulatory experts from the U.S. Nuclear Regulatory Commission (hereafter referred to as the NRC, Commission,[1] agency, or staff) updated the U.S. Fourth National Report, principally using agency information that is publicly available. This updated report follows the format of the "U.S. Third National Report for the Convention on Nuclear Safety," and is designed to be a stand-alone document. Hence, this report duplicates some of the information presented in the 2004 (third) report. To facilitate peer review, this report includes a summary table of the main changes to the report. Also new is a contribution by the Institute of Nuclear Power Operations (INPO) describing work done by the U.S. nuclear industry to ensure safety (see Part 3). INPO is a non-governmental corporation founded in 1979 by the U.S. nuclear industry to collectively promote the highest levels of safety and reliability of U. S. nuclear plants. Since the prime responsibility for the safety of a nuclear installation rests with the license holder, it was decided to involve the nuclear industry collectively through INPO, in explaining how the nuclear industry maintains and improves nuclear safety.

---

[1]   "Commission" may also refer to the Chairman and Commissioners who head the NRC.

Following the introduction, Part I of the report continues with a summary of changes to the report (Table 1), to make it easier for contracting parties to review sections of interest to them. This table is followed by a section on the conclusions of the previous review meeting. Part 2 discusses Convention articles 6–19 and includes an annex and references to provide more detailed information as appropriate. Chapters are numbered according to the article of the Convention under consideration. Each chapter begins with the text of the article, followed by an overview of the material covered by the chapter, and a discussion of how the United States meets the obligations of the article. Articles 6–9 summarize the legislative and regulatory system governing the safety of nuclear installations and discuss the adequacy and effectiveness of that system. Articles 10–16 address general safety considerations and summarize major safety-related features. Articles 17–19 address the safety of installations. Part 2 concludes with a series of appendices that discuss the NRC's main challenges as described in the NRC Strategic Plan and the 2006 Inspector General's report, followed by appendices of references, abbreviations, and acknowledgments. Annex 1 of the report is a list of U. S. nuclear plants. Part 3 of the report is a new section that contains INPO's contribution.

This report does not explicitly discuss Articles 1–5. These articles are fulfilled throughout the report, and by the existence of the report, itself. In accordance with Article 1, the report illustrates how the U.S. Government meets the objectives of the Convention. The report also discusses the safety of nuclear installations according to the definition in Article 2 and the scope of Article 3. It addresses implementing measures (such as national laws, legislation, regulations, and administrative means) according to Article 4. Submission of this report fulfills the obligation of Article 5 on reporting. In addition, the information in this report is available in more detail at the NRC's public Web site.

Summary of Changes to the Fourth U.S. National Report

To facilitate peer review of this report, the following table summarizes the changes from the third U.S. National Report. Changes from the third report are identified by a revision bar along the left margin of the page.

Table 1 Summary of Changes to the Fourth U.S. National Report

Section of Report	Change
PART 1	
INTRODUCTION	Updated.
Purpose and Structure of this Report	Added table of main changes and discussed modified structure, including the contribution by INPO.
The U.S. National Policy toward Nuclear Activities	No change.
National Nuclear Programs	
Power Uprate Program	Shortened description.
New Reactor Licensing	Updated number of design certifications and designs under review; added discussion of early site permits and of a new construction inspection organization.
Reactor Oversight Process	Updated to incorporate 2006 assessment results of the now 7-year old process.
License Renewal	Updated number of renewals. Shortened discussion.
Survey of Main Current Safety Issues	Updated status/resolution of 2004 issues and added new issues.
Reactor Materials Degradation Issues	Added new safety issues.
Unanticipated Equipment Problems from Power Uprates	Updated as safety issue.
PWR Post-LOCA Chemical Formation	Added as new safety issue.
Status of Safety Issues Discussed in Third U.S. National Report	Added.
Reactor Materials Degradation Issues	Updated status.
PWR Containment Sump Performance	Discussed above under PWR Post-LOCA Chemical Formation.
Electric Grid Reliability	Updated status.
Emergency Preparedness and Security	Updated status.
Other Major Regulatory Accomplishments	Updated to discuss early site permits, regulation involving design-basis threats, the AP1000 design certification, and risk-informed categorization and treatment of structures, systems, and components for nuclear power reactors.
Nuclear Installations in the United States (Table)	Removed because it duplicates Section 6.1.

Section of Report	Change
CONCLUSIONS ON THE U.S. NATIONAL REPORT FROM THIRD REVIEW MEETING	New.
Items Resulting from Country Group Session	New.
NRC Areas for Improvement and Major Challenges for the Future	No change in 2004–2009 NRC Strategic Plan.
NRC Major Management Challenges for the Future	Updated to incorporate 2006 Inspector General report. Eight of the nine challenges are essentially the same as in the previous U.S. National Report. In 2006, the Inspector General identified a new challenge titled, Ability to Meet the Demand for Licensing New Reactors and removed a challenge titled, Intraagency Communication.
PART 2: RESPONSE TO ARTICLES OF THE CONVENTION	
ARTICLE 6. EXISTING NUCLEAR INSTALLATIONS	
6.1 Introduction	New. Renumbered subsequent subsections.
6.2 Nuclear Installations in the United States	Updated to current status.
6.3 Regulatory Processes and Programs	
6.3.1 Reactor Licensing	Updated expectations about early site permits and design certification applications. Current power uprate issues are discussed under Survey of Main Current Safety Issues.
6.3.2 Reactor Oversight Process	Condensed section and substituted new material.
6.3.3 Industry Trends Program	Updated current experience.
6.3.4 Accident Sequence Precursor Program	Updated current experience and added some details on changes in the program.
6.3.5 Operating Experience Program	Added.
6.3.6 Program for Resolving Generic Issues	Updated NUREG-0933 to reference current revision (October 2006).
6.3.7 Rulemaking	No change.
6.3.8 Fire Regulation Program	Added.
6.3.9 Decommissioning	Shortened description and added early consideration of decommissioning.
6.3.10 Research Program	Added work regarding new reactors.
6.3.11 Programs for Public Participation, Handling Petitions, and Allegations	Moved discussion of Differing Professional Opinions; made minor changes to allegation discussion. Added description of petition process for rulemaking.
ARTICLE 7. LEGISLATIVE AND REGULATORY FRAMEWORK	

Section of Report	Change
7.1 Legislative and Regulatory Framework	No change.
7.2 Provisions of the Legislative and Regulatory Framework	
7.2.1 National Safety Requirements and Regulations	Shortened discussion.
7.2.2 Licensing of Nuclear Installations	Revised discussion of Part 52.
7.2.3 Inspection and Assessment	No change.
7.2.4 Enforcement	Updated the amount adjusted by the Federal Civil Penalties Inflation Adjustment Act (currently $130,000).
ARTICLE 8. REGULATORY BODY	Shortened descriptions of offices.
8.1 The Regulatory Body	Reorganized and renumbered subsequent subsections.
8.1.1 Mandate	No change.
8.1.2 Authority and Responsibilities	Reorganized to include discussion of NRC as an independent agency.
8.1.3 Structure of the Regulatory Body	Noted organizational changes.
8.1.4 International Responsibilities and Activities	Updated international activities.
8.1.5 Financial and Human Resources	Updated budget information. Added discussion of human resources regarding workforce planning and knowledge management.
8.1.6 Position of the NRC in the Governmental Structure	Added Department of Transportation. Updated.
8.1.7 Report of the Integrated Regulatory Review Service Self-Assessment Team	New.
8.2 Separation of Functions of the Regulatory Body from Those of Bodies Promoting Nuclear Energy	Shortened.
ARTICLE 9. RESPONSIBILITY OF THE LICENSE HOLDER	
9.1 Introduction	No change.
9.2 The Licensee's Prime Responsibility for Safety	No change.
9.3 The NRC Enforcement Program	Updated debt collection dollar amount. Added Alternative Dispute Resolution Program. Updated experience.
ARTICLE 10. PRIORITY TO SAFETY	
10.1 Background	Replaced previous examples of risk-informed rules with current ones. Added discussion of proposed set of risk-informed, performance-based requirements (10 CFR Part 53).

Section of Report	Change
10.2 Probabilistic Risk Assessment Policy	Shortened.
10.3 Applications of Probabilistic Risk Assessment	Added discussion of a risk-informed, performance-based plan and deleted discussion of the Risk-Informed Regulation Implementation Plan. Deleted previous discussions and substituted with Recent Developments.
10.4 Safety Culture	
10.4.1 NRC Monitoring of Licensee Safety Culture	Moved and thoroughly revised.
10.4.2 NRC Safety Culture	Thoroughly revised.
ARTICLE 11. FINANCIAL AND HUMAN RESOURCES	
11.1 Financial Resources	Updated to incorporate changes in the dollar amounts for liability under the Price-Anderson Act of 1957. Condensed section.
11.2 Regulatory Requirements for Qualifying, Training, and Retraining Personnel	Thoroughly revised. Shortened discussions.
ARTICLE 12. HUMAN FACTORS	Updated the program on human performance, documents, issues, and experience.
12.1 Goals and Mission of the Program	Portions moved to new 12.2 and 12.3.
12.2. The NRC Program on Human Performance	Updated the program elements.
12.3 Significant Regulatory Activities	Expanded discussion of fitness-for-duty. Thoroughly revised discussion of Human Factors Information System. Updated experience.
12.2 Significant Research Activities	Deleted.
ARTICLE 13. QUALITY ASSURANCE	Condensed section. Added a discussion of the construction inspection program for expected new reactors.
ARTICLE 14. ASSESSMENT AND VERIFICATION OF SAFETY	
14.1 Ensuring Safety Assessments throughout Plant Life	Updated the Brown's Ferry restart.
14.1.1 Maintaining the Licensing Basis	Added discussion of amendment process.
14.1.2 License Renewal	Minor editorial change.
14.1.3 The United States and Periodic Safety Reviews	No change.
14.2 Verification by Analysis, Surveillance, Testing and Inspection	Added paragraph on license conditions (10 CFR 50.54f).
ARTICLE 15. RADIATION PROTECTION	

Section of Report	Change
15.1 Authorities and Principles	Updated work on 2007 ICRP recommendations.
15.2 Regulatory Framework	Minor editorial changes.
15.3 Regulations	Minor edits and clarifications.
15.4 Radiation Protection Activities	Updated dose rates; expanded the discussion of Appendix I to 10 CFR Part 20. Added clarifications.
15.4.1 Control of Radiation Exposure of Occupational Workers	Updated dose information.
15.4.2 Control of Radiation Exposure of Members of the Public	Added ground water contamination issue.
ARTICLE 16. EMERGENCY PREPAREDNESS	
16.1 Background	No change.
16.2 Offsite Emergency Planning and Preparedness	Updated section to change references to FEMA to DHS/FEMA.
16.3 Emergency Classification System and Emergency Action Levels	No change.
16.4 Recommendations for Protective Action in Severe Accidents	Added sentence on generic communications. States changes in potassium iodide considerations.
16.5 Inspection Practices—Reactor Oversight Process for Emergency Preparedness	Changed the names of areas to be inspected.
16.6 Responding to an Emergency	Discussed changes resulting from the response to the terrorist events of September 11, 2001, and Hurricane Katrina. Updated security requirements.
16.7 International Arrangements	No change.
ARTICLE 17. SITING	
17.1 Background	Updated status of early site permits and siting applications.
17.2 Safety Elements of Siting	
17.2.1 Background	Added new developments in guidance.
17.2.2 Assessments of Seismic and Geological Aspects of Siting	Added new developments in seismic hazard assessments.
17.2.3 Assessments of Radiological Consequences	Updated experience in the use of an alternative source term.
17.3 Environmental Protection Elements of Siting	
17.3.1 Governing Documents and Process	Added some details on site approval process.
17.3.2 Other Considerations for Siting Reviews	Cited updated guidance.
ARTICLE 18. DESIGN AND CONSTRUCTION	

Section of Report	Change
18.1 Defense-in-Depth Philosophy	
18.1.1 Governing Documents and Process	No change.
18.1.2 Experience	Updated the reactivation of Watts Bar and the designs certified and under review.
18.2 Technologies Proven by Experience or Qualified by Testing or Analysis	No change.
18.3 Design for Reliable, Stable, and Easily Manageable Operation	
18.3.1 Governing Documents and Process	Added governing documents.
18.3.2 Experience	Added experience, digital instrumentation and control, and cyber security.
ARTICLE 19. OPERATION	
19.1 Initial Authorization to Operate	No change.
19.1.1 Governing Documents and Process	No change. Now 19.1.
19.1.2 Experience	Deleted.
19.2 Operational Limits and Conditions Are Defined and Revised	Shortened.
19.3 Approved Procedures	Shortened.
19.4 Procedures for Responding to Anticipated Operational Occurrences and Accidents	Shortened.
19.5 Availability of Engineering and Technical Support	No change.
19.6 Incident Reporting	Expanded.
19.7 Programs to Collect and Analyze Operating Experience	Expanded and revised to discuss revised Operating Experience Program.
19.8 Radioactive Waste	Updated to cite most recent report for the Joint Convention on the Safety of Spent Fuel Management and on the Safety of Radioactive Waste Management.
Appendix A: NRC Strategic Plan 2004-2009	No change. The NRC is currently developing a new strategic plan.
Appendix B: NRC Major Management Challenges For The Future	Updated.
Appendix C: References	Updated.
Appendix D: Abbreviations	Updated.
Appendix E: Acknowledgments	Updated.
ANNEX 1: U.S. Commercial Nuclear Power Reactors	Updated.

Section of Report	Change
PART 3: ROLE OF THE INSTITUTE OF NUCLEAR POWER OPERATIONS IN SUPPORTING THE U.S. COMMERCIAL NUCLEAR ELECTRIC UTILITY INDUSTRY'S FOCUS ON NUCLEAR SAFETY	New.

The U.S. National Policy toward Nuclear Activities

The Energy Reorganization Act of 1974 created the NRC as an independent agency of the Federal Government. The agency's mission is to license and regulate the Nation's civilian use of byproduct, source, and special nuclear materials to ensure adequate protection of public health and safety, promote the common defense and security, and protect the environment. The agency also has a role in combating the proliferation of nuclear materials worldwide. The NRC's safety and security responsibilities stem from the Atomic Energy Act of 1954, as amended. The agency accomplishes its mission by licensing and overseeing nuclear reactor operations and other activities that apply to the possession of nuclear materials and wastes, ensuring that nuclear materials and facilities are safeguarded from theft and radiological sabotage, issuing rules and standards, inspecting nuclear facilities, and enforcing regulations.

The NRC views nuclear regulation as the public's business and, as such, it must be transacted openly and candidly to maintain the public's confidence. The agency's goal to ensure openness explicitly recognizes that the public must be informed about, and have a reasonable opportunity to participate meaningfully in, the regulatory process. Except for certain proprietary business material, facility safeguards information, sensitive pre-decisional information, and information supplied by foreign countries that is deemed to be sensitive, the NRC makes the documentation that it uses in its decision-making available in the agency's Public Document Room in Rockville, Maryland, and on the agency's public Web site at http://www.nrc.gov. As a result, a significant amount of information about nuclear activities and the national policy regarding them is available to everyone.

The NRC's interpretation of regulations continues to evolve from a prescriptive, deterministic approach toward a more risk-informed and performance-based regulatory approach. Improved probabilistic risk assessment (PRA) techniques, combined with more than four decades of accumulated experience with operating nuclear power reactors, led the Commission to revise or eliminate certain requirements. The Commission is also prepared to strengthen the regulatory system when risk considerations reveal the need.

National Nuclear Programs

The NRC has a number of programs and processes to protect the health and safety of the public and the environment and to meet the obligations of the Convention. Key programs and processes in the reactor arena comprise a well-established licensing process, which includes power uprates, new reactor licensing and early site permits, reactor oversight, and license renewal. For more details on these programs, see the section of this report that pertains to Article 6.

Power Uprate Program

Under its licensing program, the NRC carefully reviews requests to raise the maximum thermal power level at which a plant may be operated, known as power uprate requests. The focus of the NRC review of these requests has been, and will continue to be, on safety. The agency closely monitors operating experience to identify safety issues that may affect the implementation of power uprates.

Power uprates can be classified as (1) measurement uncertainty recapture power uprates, (2) stretch power uprates, and (3) extended power uprates (EPUs). Measurement uncertainty recapture power uprates are less than a two percent increase and are achieved by implementing enhanced techniques for calculating reactor power. Stretch power uprates are typically up to an eight percent increase and are generally within the design capacity of the plant. Stretch power uprates usually involve changes to instrumentation setpoints and do not generally involve major plant modifications. EPUs are usually greater than stretch power uprates and require significant modifications to major balance-of-plant equipment. The NRC has approved EPUs of up to 20 percent.

Power uprate issues are discussed under the heading, Survey of Main Current Safety Issues.

New Reactor Licensing

The industry has expressed interest in constructing new nuclear power plants in the United States and indicated as of June 2007, that it may submit applications for up to 28 new reactor licenses over the next few years. The NRC is ready to accept these applications. New nuclear power plants will likely use the licensing process specified in the recently revised Title 10, Part 52, "Early Site Permits; Standard Design Certifications; and Combined Licenses for Nuclear Power Plants," of the Code of Federal Regulations (10 CFR Part 52), which is designed to be more stable and predictable than the process specified in 10 CFR Part 50, "Domestic Licensing of Production and Utilization Facilities." This licensing process resolves all safety and environmental issues, as well as emergency preparedness and security, before a new nuclear power plant is constructed.

The Commission has certified four new reactor designs under 10 CFR Part 52, making them readily available for referencing in new plant licensing submittals. These designs are General Electric's Advanced Boiling-Water Reactor, and Westinghouse's AP600, AP1000, and System 80+ (designed and licensed by Combustion Engineering).

In addition to the four advanced reactor designs that have been certified, staff is currently reviewing General Electric's economic simplified boiling-water reactor (ESBWR) design certification application. Furthermore, the NRC is preparing to receive and review two additional design certification applications in the coming year.

By certifying nuclear reactor designs, the NRC resolves safety issues in a design certification rulemaking. When an applicant submits an application for construction of a new nuclear power plant using one of the certified designs, the license application review can proceed more efficiently in a manner that ensures safety while minimizing unnecessary regulatory burden and delays.

The NRC has received four early site permit applications for sites in Virginia, Illinois, Mississippi, and Georgia. These sites currently host operating reactors. In March 2007, the Commission approved the issuance of early site permits for the Illinois and Mississippi sites. These are the first early site permits issued by the NRC and the first time this part of the 10 CFR Part 52 licensing process has been implemented. According to this process, environmental issues that have been resolved in the early site permit proceedings cannot be re-opened during a combined license proceeding.

In 2006, to better prepare the agency for the anticipated new reactor licensing and construction inspection work, while ensuring that the agency maintains its focus on the safety and security of currently operating reactors, the NRC established the Office of New Reactors. The agency also established a dedicated Construction Inspection Organization in its Region II office in Atlanta, Georgia that will carry out all construction inspection activities across the U.S., including both the day-to-day onsite inspections and the specialized inspections needed to support NRC oversight of the construction of new nuclear power plants.

There is one partially built plant, Watts Bar Nuclear Plant Unit 2, that had stopped construction activities in the mid-1980s, and is planning to resume construction and pursue operating license approval under 10 CFR Part 50. Watts Bar Unit 2 is a Westinghouse designed PWR located in southeastern Tennessee and owned by the Tennessee Valley Authority.

In addition to working on domestic issues for new reactor construction, the NRC has been a leader in cooperating with other national nuclear regulatory authorities to address advanced reactor oversight. The NRC is participating in a multinational effort, the Multinational Design Evaluation Program, to more efficiently review new reactor designs. The goal of this effort is to make all new reactor reviews more safety-focused. NRC representatives are communicating closely with representatives from the Finnish and French regulatory authorities concerning the European Power Reactor designs that are under construction in Finland and slated to be licensed in France and the United States. A longer-term multinational effort is being undertaken to establish reference regulatory practices and regulations for the review of current and future reactor designs.

Reactor Oversight Process

The NRC's Reactor Oversight Process is now nearly 7 years old. In its annual self-assessment for calendar year (CY) 2006, the NRC staff concluded that the Reactor Oversight Process provided effective safety oversight. The Reactor Oversight Process is objective, risk informed, understandable, and predictable and ensures safety, openness, and effectiveness. The NRC has appropriately monitored operating nuclear power plant activities and focused agency resources on performance issues in 2006, and plants continue to receive a level of regulatory oversight commensurate with their performance. The NRC staff expects to continually improve the Reactor Oversight Process using lessons learned and recommendations from independent evaluations. The NRC will also continue to actively solicit input from its internal and external stakeholders to further improve the Reactor Oversight Process.

License Renewal

The focus of the Commission's review of license renewal applications is on maintaining plant safety, with the primary emphasis on the effects of aging on important structures, systems, and components (SSCs). The review of a renewal application proceeds along two paths—one to review safety issues and the other to assess potential environmental impacts. Applicants must demonstrate that they have identified and can manage the effects of aging and can continue to maintain an acceptable level of safety throughout the period of extended operation. Applicants must also address the environmental impacts from extended operation. With the improved economic conditions for operating nuclear power plants, the Commission has seen sustained, strong interest in license renewal, which allows plants to operate up to 20 years beyond their current operating licenses. The Atomic Energy Act established the original 40-year term, which was not based on technical limitations.

The decision to seek license renewal is voluntary and rests entirely with nuclear power plant owners. The decision is typically based on the plant's economic viability and whether it can continue to meet the Commission's requirements. If the Commission approves all of the applications that are currently under review, approximately one-half of the plants in the United States will have had their operating licenses renewed. Based on statements from industry representatives, the Commission expects nearly all sites to apply for license renewal.

Survey of Current Safety Issues

The NRC and its licensees currently face the following regulatory and safety issues:

- reactor materials degradation issues

- unanticipated equipment problems from power uprates

- pressurized-water reactor (PWR) Emergency Core Cooling System (ECCS) sump blockage due to post-loss-of-coolant accident (LOCA) chemical formation

Reactor Materials Degradation Issues

Cases involving materials degradation include the Wolf Creek pressurizer dissimilar metal butt welds, and the Duane Arnold jet pump riser safe end.

Wolf Creek Pressurizer Dissimilar Metal Butt Weld Cracking

In October 2006, Wolf Creek reported the presence of five circumferential indications in dissimilar metal butt welds for pressurizer surge, safety, and relief nozzles, which was not acceptable under Section XI of the American Society of Mechanical Engineers (ASME) Code. The indications were determined to be flaws, most likely caused by PWSCC. The weld material is Alloy 82/182, which is susceptible to PWSCC.

These flaws were of concern because they were circumferential and much larger than those that had been previously discovered. Past indications have been axial or short circumferential flaws. Also, this is the first time that multiple circumferential indications have been identified in a single weld. This calls into question the assumptions and evaluations that the industry has made in developing the timeliness guidelines for inspection of these welds.

The NRC staff analyzed the flaws in each of the three nozzle welds found at Wolf Creek to gain insights as to when the flaws may have begun and to estimate how long it would have taken for leakage or rupture to have occurred if the flaws had been left in service. All scenarios analyzed for the surge nozzle flaw indicated that there was ample time, from the onset of leakage, for the licensee to identify the leak and take appropriate action before failure of the weld. However, for the flaws identified in the relief and safety nozzles, the analyses indicated that the flaws could have caused the welds to fail in less than three years. A number of those analyses indicated that the failure could have occurred without any prior leakage, which would serve as a warning of impending failure.

15

On the basis of the above information, the NRC issued confirmatory action letters to affected licensees, asking them to confirm their commitment to inspect the pressurizer surge, spray, safety, and relief nozzle welds by December 31, 2007. The letters asked to (1) implement enhanced reactor coolant system leakage monitoring until the inspections are complete, (2) repeat butt weld examinations every four years until the welds are either removed from service or mitigated, and (3) report inspection results to the NRC.

Duane Arnold Jet Pump Riser Safe End Cracking Event

In February 2007, the licensee (Florida Power and Light) for Duane Arnold informed the NRC that ultrasonic testing of an Inconel 82/182 weld between a low-alloy steel reactor vessel nozzle and a stainless steel jet pump riser pipe safe end revealed evidence of a likely stress-corrosion cracking flaw. Ultrasonic testing of eight Inconel 82/182 welds revealed evidence of two flaws in welds between low-alloy steel reactor vessel nozzles and stainless steel jet pump riser pipe safe ends. The piping at these locations is approximately one inch in wall thickness. Both indications are located within the weld material, on the surface connected to the inside diameter of the piping, and are approximately 55–75 percent through wall. Upon reevaluating data from 1999 and 2005 ultrasonic testing examinations, the licensee determined that these flaws had been evident at the time of those examinations, but had not been not identified. The licensee repaired the welds with overlays. The plant-specific and generic implications of the Duane Arnold information are now being evaluated.

Unanticipated Equipment Problems from Power Uprates

The NRC monitors operating experience at plants that have implemented power uprates. In 2002 and 2003, steam dryer cracking and flow-induced vibration damage on components and supports for the main steam lines and feedwater lines occurred at some boiling water reactor (BWR) plants. For example, the safety-related electromatic relief valves in the main steamlines at Quad Cities Units 1 and 2 were damaged by excessive vibration during uprated power operation (plants were uprated by up to 17.8 percent). Power upratings of more than eight percent are defined by the NRC as EPUs. Also, the steam dryers at Quad Cities Units 1 and 2 developed cracks and, in some cases, fractured metal parts from the steam dryer fell into the reactor pressure vessel and entered the steamlines leading to the turbine generator during EPU operation. In addition, feedwater sampling probes at Dresden Units 2 and 3 broke loose within a relatively short period of time under the higher feedwater flow conditions (authorized to 17 percent).

At EPU conditions, steamflow velocities can increase significantly. Plant experience has shown that the higher main steamline flow can create an acoustic resonance in the steamlines as the flow passes over branch lines. The acoustic resonance can create pressure waves that strike the steam dryer in BWRs with sufficient force to cause the stress in the steam dryer to exceed the material fatigue limits. The acoustic resonance can also cause excessive vibration that damages steamline components, such as relief valves and piping.

Therefore, the NRC has been performing more detailed reviews and inspections of plant performance and power uprate license amendment requests with respect to adverse flow effects on plant SSCs. However, the recognition that acoustic resonance can cause adverse flow effects is relatively new to the nuclear power industry. The following discussion summarizes the more significant plant operating experience related to adverse flow effects.

In response to the failure of the original steam dryers at Quad Cities Units 1 and 2 during EPU operation, the licensee installed new, improved steam dryers, in May 2005. The Quad Cities Unit 2 steam dryer design included pressure sensors, strain gages, and accelerometers to monitor the loads on the steam dryer during restart to EPU conditions. The main steamlines on both units were also instrumented to monitor loads during power ascension to EPU conditions. Following the return to EPU operation in mid-2005, the licensee discovered significant unexpected degradation of the actuators for several electromatic relief valves in the main steamlines at Quad Cities Units 1 and 2. To reduce the acoustically-generated pressure fluctuations and vibrations in the main steamlines, the licensee installed acoustic side branches in the inlet lines of the electromatic relief valves and the main steam safety valves.

The original steam dryers in Dresden Units 2 and 3 were similar to the original Quad Cities steam dryers, so they were subsequently modified to increase their structural capability. The licensee initially operated the Dresden units at EPU conditions for several years without significant damage. However, after discovering steam dryer damage at the Dresden units in 2005 and 2006, the licensee replaced the steam dryer in Dresden Unit 3 in November 2006. The steam dryer in Dresden Unit 2 will be replaced in October 2007.

The NRC is applying lessons learned from operating experience and analysis of potential adverse flow effects in reviewing power uprate requests for operating nuclear power plants and design certification requests for new nuclear power plants. As part of this effort, the NRC has updated relevant sections of NUREG-0800, "Standard Review Plan for the Review of Safety Analysis Reports for Nuclear Power Plants", and Regulatory Guide 1.20, "Comprehensive Vibration Assessment Program for Reactor Internals during Preoperational and Initial Startup Testing," issued in March 2007, to further guide NRC reviewers and the nuclear industry to evaluate potential adverse flow effects when considering power uprates in both operating and new nuclear power plants.

PWR Post-LOCA Chemical Formation

The third U.S. National Report identified PWR containment sump performance as an issue. That issue remains of concern, though substantial progress has been made in resolving it (e.g., by greatly enhancing strainer size and removing problematic materials). Consistent with NRC Generic Letter (GL) 2004-02, "Potential Impact of Debris Blockage on Emergency Recirculation during Design Basis Accidents at Pressurized-Water Reactors," dated September 13, 2004, the NRC continues to target December 31, 2007, as the date for licensees to demonstrate that their strainers will perform acceptably in the presence of expected plant-specific debris generation and transport. However, one aspect of the issue, the potential for chemical effects on strainers and downstream components, has turned out to be particularly challenging.

To address concerns about the potential for chemical precipitates and corrosion products to significantly block a fiber bed and increase the head loss across an emergency core cooling system sump screen, a joint NRC/industry Integrated Chemical Effects Testing Program was started in 2004 and concluded in August 2005. The test program identified chemical precipitation products, and followup testing and analyses were performed to address the effect on head loss. On the basis of this effort, the NRC issued Information Notice (IN) 2005-26, "Results of Chemical Effects Head Loss Tests in a Simulated PWR Sump Pool Environment," on September 16, 2005.

17

The NRC conducted additional research to support evaluation efforts and provide confirmatory information. This work includes research on (1) chemical effects to determine whether the PWR sump pool environment generates byproducts that contribute to sump clogging, (2) pump head losses caused by the accumulation of containment materials and chemical byproducts, and (3) the prediction of the chemical species that may form in these environments. Supplement 1 to IN 2005-26, "Additional Results of Chemical Effects Tests in a Simulated PWR Sump Pool Environment," dated January 20, 2006, provides additional information on test results related to chemical effects in environments containing dissolved phosphate (e.g., from trisodium phosphate) and dissolved calcium. The results discussed in the information notices clearly demonstrate that chemical effects can be significant. Follow-on testing sponsored by the industry has also shown that these chemical effects tend to increase the potential for substantial head loss.

The agency also conducted research on the transportability of coating chips in containment pool environments and on the effect of ingested debris on downstream valve performance. In 2006, the staff completed additional research on various aspects of sump clogging. All planned NRC-sponsored research activities related to PWR sump clogging are now complete and documented, though information that is obtained as the staff continues its reviews of industry activities to close the issue may suggest the need for more NRC-sponsored research.

The NRC is currently reviewing an industry topical report that supports evaluation and testing of chemical effects, and expects to receive another topical report addressing chemical effects inside the reactor vessel. Licensees are developing protocols for integrated head-loss testing that will include chemical effects. The NRC is monitoring this work, reviewing the protocols, and observing and commenting on testing. Most of the testing is expected to occur in summer and fall 2007. The staff will consider the test results in the licensees' final strainer designs, and in demonstrations that their strainer designs are adequate. The NRC will use inputs from its review of licensee responses to GL 2004-02, inspections, and audits of corrective actions by selected licensees to support closing the PWR sump clogging issue, including the chemical effects.

Status of Safety Issues Discussed in Third U.S. National Report

Reactor Materials Degradation Issues

The reactor materials degradation issues in 2004 focused on PWSCC in PWR vessel upper and lower head penetrations and other locations in the reactor coolant system and boric acid corrosion. Cases discussed were a significant cavity in the reactor vessel head at the Davis-Besse Nuclear Power Station and boron deposits around lower reactor vessel head penetration nozzles at the South Texas Project Unit 1.

Davis-Besse Nuclear Power Station Reactor Vessel Head

In March 2002, the licensee for Davis-Besse Nuclear Power Station discovered a significant cavity in the reactor vessel head. The cavity was next to a leaking nozzle with a through-wall crack and in an area of the vessel head that had been covered with boric acid deposits for several years. The NRC considers the Davis-Besse case as one of the most significant in the agency's recent history. Davis-Besse continues to draw much interest and comment from the agency, the industry, all levels of government, and the public. Consequently, the NRC continues to dedicate substantial effort to resolving both the technical and programmatic issues that contributed to the degradation at this site. (For history and details, see the extensive documentation regarding this case on the NRC public Web site at www.nrc.gov.)

South Texas Instrumentation Penetrations

Another significant case concerns lower reactor vessel head penetration nozzles. The lower head and bottom-mounted instrumentation penetrations of the South Texas Project, Unit 1, reactor vessel were visually inspected on April 12, 2003, as a routine part of the unit's refueling outage. The inspection found small amounts of white residue around 2 of the 58 penetrations at the junction where the penetrations meet the lower reactor vessel head. The licensee performed destructive examination of one of the penetrations and found a through-wall flaw resulting from primary water stress-corrosion cracking (PWSCC). As a result of these events, the NRC staff issued Bulletin 2003-02, "Leakage from Reactor Pressure Vessel Lower Head Penetrations and Reactor Coolant Pressure Boundary Integrity," on August 21, 2003.

In September 2006, the NRC staff issued NUREG-1863, "Review of Responses to NRC Bulletin 2003-02—Leakage from Reactor Pressure Vessel Lower Head Penetrations and Reactor Coolant Pressure Boundary Integrity." The NUREG summarizes the available information on the current and future inspection programs and the industry and agency's ongoing plans to ensure the integrity of the penetrations. Even though there was no evidence of bottom-mounted instrumentation penetration leakage at other plants, the NRC staff believes that continuous monitoring of the penetrations is necessary to ensure the integrity of the lower reactor vessel head.

PWR Containment Sump Performance

For the status of this issue, see the discussion above under the heading, PWR Post-LOCA Chemical Formation.

Electric Grid Reliability

The long duration of the blackout in the eastern United States and Canada on August 14, 2003, highlighted the need to further consider the impact of grid reliability on nuclear plants. The blackout affected more than 290 commercial power facilities, including ten nuclear plants.

The NRC provided extensive technical support to the joint U.S.-Canada Power System Outage Task Force to evaluate nuclear power plant responses to the blackout. In April 2004, the NRC issued a regulatory information summary to inform licensees of nuclear power plants that grid reliability can impact plant risk and the availability of offsite power.

On February 1, 2006, the NRC issued GL 2006-02, "Grid Reliability and the Impact on Plant Risk and the Operability of Offsite Power." The purpose of this generic letter was to determine whether licensees were complying with NRC regulatory requirements governing electric power sources and associated personnel training. The staff did not identify any safety concerns or compliance issues as a result of its review of GL 2006-02.

In addition to the efforts described above and because of the importance of the offsite power system, the NRC continues to monitor grid reliability and works with the organizations responsible for regulating the grid (i.e., the Federal Energy Regulatory Commission, the U.S. Department of Energy (DOE), and the Electric Reliability Organization (i.e., the North American Electric Reliability Corporation)) and licensees to ensure that they are prepared to address changes in the overall reliability of the offsite power system.

Emergency Preparedness and Security

Emergency Preparedness

After the terrorist attacks on September 11, 2001, the NRC required nuclear facility licensees to assess the potential impact of a terrorist-initiated event on the site emergency plan. This additional requirement complements the emergency plan and provides assurance that licensees are prepared to respond to a terrorist event.

The NRC is continuing to develop emergency preparedness policies, regulations, programs, and guidelines for both currently licensed nuclear reactors and potential new nuclear reactors; provide technical expertise on emergency preparedness issues; coordinate, as appropriate, with other parts of the NRC and stakeholders; and oversee and provide technical direction for the Emergency Preparedness cornerstone of the Reactor Oversight Process.

Security

The NRC ordered new security-related requirements after September 11, 2001. The NRC is currently codifying many of the additional requirements imposed on licensees by those orders, based on experience and comprehensive reviews. For example, the NRC has proposed a comprehensive revision to the requirements for physical protection at nuclear reactors and published a proposed rule to amend its regulations pertaining to the design basis threat.

After September 11, 2001, the NRC issued an order (Interim Compensatory Measures Order EA-02) as part of a comprehensive effort to improve the capabilities of commercial nuclear reactor facilities to respond to terrorist threats. Section B.5.b. of the Order required licensees to develop specific guidance and strategies to maintain or restore core cooling, containment, and spent fuel pool cooling capabilities using existing or readily available resources that could be effectively implemented should the loss of large areas of the plant due to explosions or fire occur, including those that an aircraft impact might create. Although it was recognized before September 11, 2001, that nuclear reactors already had significant capabilities to withstand a broad range of attacks, implementing these mitigation strategies would significantly enhance the plants' capabilities to withstand a broad range of threats.

Licensee have been implementing Section B.5.b mitigation strategies since the Order was issued. In 2005, the NRC issued guidance to more fully describe its expectations for implementing Section B.5.b. The guidance relied upon lessons learned from detailed NRC engineering studies and industry best practices. Additionally, the NRC conducted two on-site team assessments at each reactor facility that identified additional mitigating strategies to preserve core cooling, containment integrity, and spent fuel pool cooling. The NRC has incorporated requirements for the B.5.b. mitigating strategies into facilities' operating licenses. These strategies include fire-fighting response strategies, operations to mitigate fuel damage, and actions to minimize release of radioactive material. Elements that can be part of these strategies include an assessment of mutual aid fire fighting assets, developing designated staging areas for the equipment and materials, communications protocols, training on integrated fire response strategy, and water spray scrubbing. In total, these enhancements strengthen the interface between plant safety and security operations and add to the safety approach of defense-in-depth.

Since September 11, 2001, the issue of an airborne attack on U.S. infrastructure, including both operating and potential new nuclear power plants, has been widely discussed. The NRC has comprehensively studied the effect of an airborne attack on nuclear power plants. Studies confirm that there is a low likelihood that an airplane attack on a nuclear power plant would affect public health and safety, owing in part to the inherent robustness of the structures. One study identified new methods plants could use to minimize damage and risk to the public in the event of any kind of large fire or explosion. Nuclear power plants subsequently implemented many of these methods. The NRC is now considering new regulations for future reactors' security regarding inherent safety and security features to minimize potential damage from an airborne attack.

The third review meeting concluded that one NRC challenge was to ensure that the post-September 11, 2001 emphasis on security does not adversely affect operational safety. One way the NRC is meeting this challenge is by proposing to amend 10 CFR 73, "Physical Protection of Plants and Materials," to explicitly require licensees to assess and manage the safety/security interface at commercial power reactors. The proposed amendment will require (1) licensees to assess and manage the potential for adverse effects between safety and security (including the site emergency plan) before implementing changes to plant configurations, facility conditions, or security; (2) changes to be assessed and managed must include planned and emergent activities such as physical modifications, procedural changes, and maintenance activities; (3) that upon identifying potential adverse interactions, licensees must communicate them to the appropriate licensee personnel and take corrective or compensatory actions to maintain safety and security in accordance with applicable regulations, orders, and license conditions. In addition, the NRC has prepared a draft regulatory guide, DG-5021, "Managing the Safety/Security Interface," to describe acceptable methods to implement the proposed amendment, explain techniques that the NRC will use in evaluating compliance, and guide applicants and licensees.

Other Major Regulatory Accomplishments

Since its previous U.S. National Report in 2004, the NRC has issued two early site permits and amended its regulations concerning the design-basis threat (DBT), the AP1000 design certification, and risk-informed categorization and treatment of SSCs for nuclear power reactors.

Issuance of Early Site Permits for the Clinton and Grand Gulf Sites

On March 15 and April 5, 2007, the NRC issued the first early site permits to the Exelon Generation Co. for the Clinton site near Clinton, Illinois, and to System Energy Resources (SERI) for the Grand Gulf site near Port Gibson, Mississippi, respectively. The main advantage of the early site permit process is the removal of environmental contentions later in the licensing process. Successful completion of the early site permit process resolves many site-related safety and environmental issues and determines that the sites are suitable for possible future construction and operation of a nuclear power plant. The permits are valid for up to 20 years. An early site permit may be referenced in an application to the NRC for a combined license to build one or more nuclear plants on the permitted site.

Design-Basis Threat

On March 19, 2007, the NRC amended the regulations at 10 CFR Part 73, "Physical Protection of Plants and Materials," that govern DBTs. This final rule makes security requirements that are similar to those previously imposed by the Commission's April 29, 2003 DBT orders, which are generically applicable to all plants and redefine the level of security requirements necessary to ensure that the public health and safety, common defense, and security are adequately protected. The rule revises the DBT requirements for radiological sabotage, generally applicable to power reactors and Category I fuel cycle facilities, and for theft or diversion of NRC-licensed strategic special nuclear material, also applicable to Category I fuel cycle facilities.

AP1000 Design Certification

On January 27, 2006, the NRC amended 10 CFR Part 52 to certify Westinghouse's AP1000 standard plant design. This action was necessary so that applicants intending to construct and operate an AP1000 design may do so by referencing the regulation (AP1000 Design Certification Rule).

Risk-Informed Categorization and Treatment of SSCs for Nuclear Power Reactors

On November 22, 2004, the NRC amended 10 CFR Part 50 to provide an alternative approach for establishing the requirements for treatment of SSCs for nuclear power reactors using a risk-informed method of categorizing SSCs according to their safety significance. The 10 CFR 50.69 rule revises requirements with respect to "special treatment," that is, those requirements that provide increased assurance (beyond normal industrial practices) that SSCs perform their design-basis functions. This amendment permits licensees and applicants for licenses to remove SSCs of low safety significance from the scope of certain identified special treatment requirements and revises requirements for SSCs of greater safety significance. In addition to the rulemaking and its associated analyses, the Commission also issued a regulatory guide to implement the rule.

CONCLUSIONS ON THE U.S. NATIONAL REPORT
FROM THE THIRD REVIEW MEETING

This section presents the conclusions from the review of the 2004 U.S. National Report at the third review meeting in April 2005.

Delegates from other countries noted that the United States prepared and delivered a highly informative and candid presentation at the country group meeting. The country group also noted that the United States provided extensive and comprehensive answers to all of the 266 questions posted by other contracting parties.

Items Resulting from Country Group Session

Review of the questions raised by other contracting parties on the U.S. National Report identified the following areas of interest:

(1)   risk-informed regulation,
(2)   licensing and oversight,
(3)   long-term operation,
(4)   safety culture and human factors,
(5)   operating experience,
(6)   organizational structure,
(7)   emergency preparedness and security, and
(8)   radiation protection.

The NRC's presentation to Country Group 1 focused on these topics.

Country Group I participants concluded that the U.S. good practices were:

(1)   the transition toward risk-informed approaches in an incentive-based manner,
(2)   the successful implementation of the Reactor Oversight Process,
(3)   the NRC's openness as demonstrated by its Strategic Plan, improved Web site, and commitment to stakeholder involvement, and
(4)   the NRC's comprehensive and innovative strategies to attract and retain staff.

Among NRC's challenges, Country Group I identified:

(1)   ensuring that the post-September 11, 2001 emphasis on security does not have an adverse effect on operational safety,
(2)   continuing the focus on safe long-term operation and emerging technical issues,
(3)   preparing for the potential licensing of new reactors,
(4)   appropriately emphasizing safety culture, safety management, and organizational aspects of nuclear safety,
(5)   continued monitoring and analysis of operating experience and timely implementation of lessons learned, and
(6)   expanding risk-informed, performance-based concepts while ensuring a consistent regulatory approach.

The group's highlighted planned U.S. initiatives included:

(1)   enhancing the Reactor Oversight Process to increase NRC inspector review of licensee safety-conscious work environment,

(2)   working with the international community to define the ways and means to detect deteriorating performance, including safety culture and safety management, and

(3)   planning for the performance of an International Regulatory Review Team (IRRT) self-assessment (since the time of the previous report, this group has been renamed the Integrated Regulatory Review Service (IRRS)).

The current U.S. National Report addresses many of these issues under the relevant articles. Specifically, Article 10 discusses risk-informed regulation, Article 6 presents the Reactor Oversight Process, Article 14 addresses license renewal and periodic safety reviews, and Articles 17 and 18 cover new reactor licensing. Article 8 discusses the IRRS.

NRC Areas for Improvement and Major Challenges for the Future

The NRC identified major challenges for the future in its Strategic Plan 2004–2009, which were unchanged from those identified in the third U.S. National Report. Challenges arise from the changing regulatory environment and external factors. The challenges are summarized below; Appendix A to this report provides more details.

- The NRC will strengthen the interrelationship among safety, security, and emergency preparedness.

- The majority of operating nuclear power plants will have applied for license renewal to help meet the country's demand for energy. A challenge is to monitor, manage, and control the effects of aging so that safety is ensured for the renewal period.

- The U.S. Department of Energy (DOE) will apply to construct and operate the country's high-level radioactive waste repository. The timing of this action may challenge the allocation of the NRC's resources.

- The U.S. nuclear power industry will show a growing interest in licensing and constructing new nuclear power plants to meet the Nation's demand for energy. Challenges include analyzing in detail the vulnerability to accidents and security compromises, as well as developing appropriate inspections, tests, analyses, and acceptance criteria for construction.

- The NRC will continue to see increased requirements to coordinate with a wide array of Federal, State, and local agencies related to homeland security and emergency planning.

- The regulatory climate is expected to adjust to both internal and external factors (described below).

Key external factors that could cause challenges are:

- Receipt of new reactor operating license applications

- A significant operating incident (domestic or international)

- A significant terrorist incident

- Timing of the DOE application and related activities for the high-level waste repository at Yucca Mountain

- Homeland security initiatives

- Legislative initiatives

NRC Major Management Challenges for the Future

By law, the Inspector General of each Federal agency (discussed in Article 8) identifies the agency's most serious management and performance challenges facing the agency and assesses progress in addressing those challenges. The NRC's Inspector General's annual assessment of the major management challenges confronting the agency can be found on NRC's public Web site. Eight of the nine challenges are essentially the same as those identified in the previous U.S. National Report. In 2006, the Inspector General identified a new challenge titled, Ability to Meet the Demand for Licensing New Reactors, and removed a challenge titled, Intraagency Communication. The main challenges are summarized below. For more detail, see Appendix B to this report.

- Protection of nuclear material used for civilian purposes

- Protection of information

- Development and implementation of a risk-informed and performance-based regulatory oversight approach

- Ability to modify regulatory processes to meet a changing environment

- Implementation of information resources

- Administration of all aspects of financial management

- Communication with external stakeholders throughout the NRC's regulatory activities

- Managing human capital

- Ability to meet the demand for licensing new reactors

PART 2

ARTICLE 6. EXISTING NUCLEAR INSTALLATIONS

Each Contracting Party shall take the appropriate steps to ensure that the safety of nuclear installations existing at the time the Convention enters into force for that Contracting Party is reviewed as soon as possible. When necessary in the context of this Convention, the Contracting Party shall ensure that all reasonable practicable improvements are made as a matter of urgency to upgrade the safety of the nuclear installation. If such upgrading cannot be achieved, plans should be implemented to shut down the nuclear installation as soon as practically possible. The timing of the shutdown may take into account the whole energy context and possible alternatives, as well as the social, environmental, and economic impact.

This section explains how the United States ensures the safety of nuclear installations in accordance with the obligations in Article 6. This section covers the reactor licensing and major oversight processes in the United States. This section also discusses programs for rulemaking, fire protection regulation, decommissioning, research, and programs for public participation. The NRC posts the major results of assessments on the agency's public Web site at http://www.nrc.gov. This update includes expectations about early site permits and design certification applications, current experience, and revised details about programs.

6.1 Introduction

The NRC's primary goal is safety. The Agency achieves this goal by ensuring that the performance of licensees is at or above acceptable safety levels. The NRC's licensees are responsible for designing, constructing, and operating nuclear facilities safely, while the NRC is responsible for the regulatory oversight of the licensees. Five strategic outcomes for this goal are specified:

- No nuclear reactor accidents.
- No inadvertent criticality events.
- No acute radiation exposures resulting in fatalities.
- No releases of radioactive materials that result in significant radiation exposures.
- No releases of radioactive materials that cause significant adverse environmental impacts.

The NRC met all of its safety strategic outcomes in fiscal years (FYs) 2005 and 2006.

The NRC also uses performance measures to determine whether the Agency has met its safety goal. The NRC met its performance measures in FYs 2005 and 2006. Currently the NRC uses six performance measures.

The first measure analyzes plant performance based on a large number of performance indicators and inspection findings.

The second measure tracks significant precursor events determined by the likelihood of an event adversely impacting safety.

The third performance measure indicates whether the NRC identifies significant issues in a plant during inspections conducted under the reactor oversight program.

The fourth measure tracks the trends of several key indicators of nuclear power plant safety. This measure is the broadest measure of the safety of nuclear power plants, incorporating the performance results from all plants to determine industry average results.

These measures indicated that not only were the plants safely operated, but the events that did occur were of relatively minor significance.

The other two measures address harmful radiation exposures to the public and occupational workers and radiation exposures that harm the environment. Neither of these measures exceeded their targets in FY 2006.

6.2 Nuclear Installations in the United States

For a list of all 104 nuclear installations in the United States, see Annex 1. The source of this list is the NRC Information Digest, 2006-2007 edition, available on the agency's Web site.

6.3 Regulatory Processes and Programs

6.3.1 Reactor Licensing

To construct and operate a nuclear reactor, an entity must submit an application to the NRC for a safety and environmental review. The public has opportunities to participate through a hearing process. The NRC licensed all current operating nuclear plants under the detailed two-step process specified in 10 CFR Part 50, first issuing a construction permit and then an operating license. Although the NRC has not received any applications for reactor licenses since 1976, it has recently received applications for early site permits and design certifications. In addition, due largely to the favorable incentives created by the U.S. Congress in the Energy Policy Act of 2005, the industry has indicated that it will be submitting applications for two additional design certifications, and potentially as many as 20 applications for up to 28 reactor licenses during 2007 through 2009. The agency will review these applications under a new licensing process, specified in 10 CFR Part 52. Article 18 provides more detail about the 10 CFR Part 52 regulations.

The NRC's reactor licensing process provides for the review and approval of changes after initial licensing. The process allows amendments to the operating license to support plant changes, license renewal, changes of ownership and license transfer, exemptions and relief from NRC regulations, and increasing the reactor power level (i.e., power uprates). The process is discussed further in the Introduction and in other articles (Articles 14, 17 and 18).

6.3.2 Reactor Oversight Process

Through its Reactor Oversight Process the NRC continuously oversees nuclear power plants to verify that they are being operated in accordance with the agency's rules and regulations. NRC has full authority to take whatever action is necessary to protect public health and safety, and may demand immediate licensee actions, up to and including a plant shutdown.

The Reactor Oversight Process uses both inspection findings and performance indicators to assess the performance of each plant within a regulatory framework of seven cornerstones of safety. Toward that end, the NRC performs a program of baseline inspections at each plant and may perform supplemental inspections and take additional actions to ensure that the plants address significant issues. The NRC communicates the results of its oversight process by posting plant-specific inspection findings and performance indicator information on the NRC's public Web site. The NRC also conducts public meetings with licensees to discuss the results of the NRC's assessments of licensee performance.

NRC assesses the Reactor Oversight Process annually and evaluates the overall effectiveness of the Reactor Oversight Process through its success in meeting its preestablished goals (i.e., performance metrics) and intended outcomes. The latest report, SECY-07-0069, "Calendar Year 2006 Reactor Oversight Process Self-Assessment," was issued on April 6, 2007.

The calendar year 2006 self-assessment results indicated that the Reactor Oversight Process was successful in meeting its program goals and achieving its intended outcomes. The Reactor Oversight Process was deemed to be objective, risk-informed, understandable, and predictable, and met the agency goals of ensuring safety, openness, and effectiveness. The NRC staff maintained its focus on stakeholder involvement and continued to improve various aspects of the Reactor Oversight Process as a result of feedback and lessons learned. The NRC staff implemented several process improvements in 2006 to address issues raised by the Commission, recommended by independent reviews, and obtained from internal and external stakeholder feedback. Most notably, the NRC staff made significant enhancements to the assessment and inspection programs in 2006 to more fully address safety culture.

The NRC continued to improve the performance indicator program to better identify declining plant performance in a timely manner, but recognized the need for further improvement. The inspection program independently verified that plants were operated safely, appropriately identified performance issues, and ensured the adequacy of licensee corrective actions to address the noted performance issues. Further improvements in the significance determination process resulted in the timeliness goal for making final determinations being met in 2006 for the first time since implementation. [The significance determination process uses risk insights to help NRC inspectors and staff determine the safety significance of inspection findings.] The assessment program ensured the NRC and licensees took necessary actions to address identified performance issues. The NRC continues to actively solicit input from the NRC's internal and external stakeholders and expects further process improvements based on stakeholder feedback and lessons learned.

Based on its calendar year 2006 self-assessment, the NRC intends to focus on the following significant actions or activities to improve the efficiency and effectiveness of the Reactor Oversight Process in 2007: (1) continue to monitor implementation of the Mitigating Systems Performance Index and address additional improvements to the performance indicator program to better identify those plants with declining performance; (2) implement the Reactor Oversight Process realignment process and adjust inspection resources accordingly; (3) continue to monitor implementation of the safety culture enhancements and address related recommendations in this area; and (4) implement adjustments to the process to incorporate pending Commission direction related to the point at which licensee senior management will be requested to meet with the Commission to discuss actions being taken to improve performance.

6.3.3 Industry Trends Program

The NRC staff implemented the Industry Trends Program in 2001 and is continuing to develop the program as a means to confirm that the nuclear power industry is maintaining the safety of operating power plants and to increase public confidence in the effectiveness of the NRC's processes. The agency uses industry-level indicators to identify adverse trends in performance. After assessing industry trends for safety significance, the NRC responds as necessary to any identified safety issues, including adjusting the inspection and licensing programs if necessary. One important output of the Industry Trends Program is the annual agency performance measures reported to Congress on the number of statistically significant adverse industry trends. The NRC Performance and Accountability Report includes this outcome measure.

In addition to long-term trending of the data to identify statistically significant adverse trends, the NRC staff uses a statistical approach based on prediction limits to identify potential short-term, year-to-year emergent issues before they become long-term trends. Short-term trending of the fiscal year 2006 data did not identify any issues that warranted additional analysis or significant adjustments to the nuclear reactor safety inspection or licensing programs.

The Reactor Oversight Process uses both plant-level performance indicators and inspections to provide plant-specific oversight of safety performance, whereas the Industry Trends Program provides a means to assess overall industry performance using industry-level indicators. The NRC evaluates issues that are identified from either program using information from agency databases and addresses those determined to have generic safety significance, including generic safety inspections under the Reactor Oversight Process, the generic communications process, and the generic safety issue process.

Based on the information currently available from the industry-level indicators and the Accident Sequence Precursor Program (below), no statistically significant adverse industry trends have been identified through FY 2006. For more details, see SECY-07-0063, "FY 2006 Results of the Industry Trends Program for Operating Power Reactors and Status of Ongoing Development," dated April 3, 2007, which is available on the NRC public Web site.

6.3.4 Accident Sequence Precursor Program

The Accident Sequence Precursor Program systematically evaluates U.S. nuclear power plant operating experience to identify, document, and rank the operating events that are most likely to have led to inadequate core cooling and severe core damage (precursors), accounting for the likelihood of additional failures. The objectives of the program include the following:

(1)   provide a comprehensive, risk-informed view of nuclear power plant operating experience and a measure for trending nuclear power plant core damage risk

(2)   provide a partial check on dominant core damage scenarios predicted by PRAs

(3)   provide feedback for regulatory activities

To identify potential precursors, the NRC reviews plant events from licensee event reports, NRC inspection reports, and special requests from NRC staff. The NRC then analyzes any identified potential precursor by calculating a probability of an event leading to a core damage state. The program considers a precursor to be an event with a conditional core damage probability (CCDP) greater than or equal to 1×10^{-6} and a significant precursor as an event with a CCDP greater than or equal to 1×10^{-3}.

In 2006, the NRC streamlined the analysis and review processes to improve the timeliness and reduce the number of Accident Sequence Precursor Program analyses that undergo formal peer review. The analysis process now includes results from the significance determination process and the NRC Incident Investigation Program when practicable.

The staff completed precursor trend analyses as part of the annual Accident Sequence Precursor Program status report provided to the Commission in SECY-06-0208, "Status of the Accident Sequence Precursor Program and the Development of Standardized Plant Analysis Risk Models," dated October 5, 2006. The following are a few of the insights provided in the report.

(1)   The staff identified no significant precursors in FY 2005 and FY 2006. In addition, the staff identified no statistically significant trends for significant precursors during the FY 1996–2005 period.

(2)   The risk contribution from precursors was generally constant during the FY 1996–2003 period and has decreased during FY 2003-2005.

(3)   The staff evaluated precursor data through FY 2005 to identify statistically significant adverse trends for the Industry Trends Program. The staff divided the data for the rates of occurrence of all precursors during the FY 1996–2005 period into two statistically distinct groups, FY 1996–2000 and FY 2001–2005, because of program scope changes that occurred in FY 2000. The analysis detected no statistically significant trend for either of these 5-year periods.

(4)   Over the more recent time period from FY 2001–2005, the analysis detected a statistically significant decreasing trend for higher risk precursors (i.e., CCDP $\geq10^{-4}$).

6.3.5 Operating Experience Program

The NRC launched a revised Operating Experience Program in January 2005, recognizing that the effective use of operating experience is important for the safety mission of the agency. Under the current NRC Strategic Plan, the agency is committed to "evaluate and utilize domestic and international operational experience and events to enhance decision-making," as part of its effort to achieve the goal of safety. As a result, the NRC's emphasis on the effective use of operating experience remains strong.

The fundamental aim of the Operating Experience Program is to collect, evaluate, communicate, and apply operating experience information to achieve the NRC's principal safety mission of protecting people and the environment. Operating experience is reported to the NRC or identified in licensee event notifications, and in many other reports that are submitted under licensee reporting requirements of the regulations, and in reports of operating experience at foreign facilities. Sources of foreign operating experience include International Nuclear Event Scale (INES) events and Incident Reporting System reports. NRC staff systematically screens nuclear reactor related operating experience for safety significance and generic implications.

35

The NRC staff also determines the need for further action and application of lessons learned related to plant operating experience.

To support its safety mission, the NRC increased resources dedicated to operating experience review and instituted a clearinghouse. The clearinghouse collects, stores, screens, and communicates operating experience; conducts and coordinates the evaluation of operating experience; tracks the application of operating experience lessons learned; and coordinates the NRC operating experience activities with other organizations performing related functions.

Upon launching the program, the NRC developed an internal Web site to provide a centralized source for accessing reactor operating experience information. This Web site is a gateway to the agency's operating experience document collections, contacts, search tools, sources, and reference material. In addition, the NRC created an operating experience community forum to quickly disseminate operating experience to the appropriate technical staff. All of the NRC's event-related reports can be found on the agency's public Web site.

For more information on this program, see Section 19.7 of this report.

6.3.6 Program for Resolving Generic Issues

The program for resolving generic issues provides an effective means for addressing issues that are not sufficiently addressed by existing rules, guidance, or programs and that affect licensees (and others under NRC jurisdiction). The NRC maintains a complete list of all generic issues in NUREG-0933, "A Prioritization of Generic Safety Issues." The most recent revision, published in October 2006, is available on the agency's public Web site.

Sources of candidate generic issues include safety evaluations, operational events, and suggestions from individual staff members, outside organizations, or members of the general public. Existing programs generally address emergent issues that demand immediate attention (e.g., issues that may require plant shutdown) so that quick decisions can be made.

6.3.7 Rulemaking

The NRC's regulations, also called rules, impose requirements that licensees must meet to obtain or retain a license or certificate to use nuclear materials or to operate a nuclear facility. The technical staff usually proposes a rule or a change to a rule because of a perceived need to protect the public health and safety. However, any member of the public may petition the NRC to develop, change, or rescind a rule. The impetus for a proposed rule could be a requirement issued by the Commission, a petition for rulemaking submitted by a member of the public, or research results that indicate a need for a rule change. The NRC publishes the proposed rule in the *Federal Register* for public comment. Once the public comment period has closed, the staff analyzes the comments, makes any needed changes, and forwards the final rule for approval, signature, and publication in the *Federal Register*.

The NRC developed RuleForum to provide an easy means for members of the public to access and comment on NRC rulemaking actions. Accessible through NRC's public Web site, RuleForum contains proposed rulemakings that have been published in the *Federal Register*, petitions for rulemaking, and other types of documents related to rulemaking proceedings.

The NRC Commissioners must approve each final rule that involves significant matters of policy. Once approved, the final rule is published in the *Federal Register* and usually becomes effective 30 days later. The introduction to this report summarizes the significant nuclear reactor-related rules issued since the previous U.S. National Report.

6.3.8 Fire Regulation Program

The NRC has three main foci in fire protection regulation. These foci are: implementation of the new risk-informed, performance-based fire protection licensing basis (10 CFR 50.48(c)); resolution of the fire-induced multiple spurious operation/circuit analysis issue; and resolution of licensees' non-conforming post-fire operator manual actions. To support the implementation of 10 CFR 50.48(c), the NRC issued Regulatory Guide 1.205 "Risk-Informed, Performance-Based Fire Protection for Existing Light-Water Nuclear Power Plants," in May 2006. As of September 2007, there are 42 reactor units committed to transitioning to 10 CFR 50.48(c). Two nuclear stations, Oconee and Shearon Harris, volunteered to be pilot plants for the transition, and the pilots' license amendment requests are expected in May and June 2008. Regarding circuit analysis actuations, the NRC continues to engage with stakeholders to develop an acceptable method to resolve the issue. On the topic of operator manual actions, the NRC has issued Regulatory Issue Summary 2006-10 "Regulatory Expectations with Appendix R Paragraph III.G.2 Operator Manual Actions," and developed NUREG-1852 "Demonstrating the Feasibility and Reliability of Operator Manual Actions in Response to Fire." Licensees are to resolve issues involving post-fire operator manual actions by March 2009.

The NRC has an active fire research program that develops the technical bases for ongoing and future regulatory activities in fire protection and fire risk analysis. NRC currently has ongoing research in (1) developing and improving fire risk analysis methods and tools; (2) applying these methods and tools to develop risk insights; (3) collecting, generating and analyzing fire related data; (5) verifying, validating and improving fire models for regulatory use; (6) performing specialized fire testing on electrical cables for both hot shorts and fire properties; (7) evaluating shipping casks for beyond design basis fire conditions; and (8) evaluating methods to predict operator performance during fire conditions.

The fire research program supports the Agency's strategic goals of safety and effectiveness and partners with other organizations with similar missions such the National Institute of Standards and Technology, the Electric Power Research Institute (EPRI) and on the international level with groups such as Organization for Economic and Cooperation and Development Committee on the Safety of Nuclear Installations.

6.3.9 Decommissioning

The decommissioning process consists of a series of integrated activities, beginning with the nuclear facility transitioning from "active" to "decommissioning" status and concluding with termination of the license, and release of the site. The NRC has adopted extensive regulations to ensure that decommissioning is accomplished safely and that residual radioactivity is reduced to a level that permits release of the property for unrestricted use. The NRC reviews and approves license termination plans, conducts inspections, processes license amendments, and monitors the status of activities to ensure that radioactive contamination is reduced or stabilized.

The design criteria for new facility construction at 10 CFR 20.1406, "Minimization of Contamination," address early consideration of decommissioning. Furthermore, the safety standards on decommissioning promulgated by the International Atomic Energy Agency (IAEA) cite considerations, which the United States supports, for future decommissioning provisions in the conceptual design of nuclear facilities.

Spent fuel can remain stored in the spent fuel pool or in dry cask storage facilities until a geologic repository is built and operating. The NRC regulations in 10 CFR Part 50 and 10 CFR Part 72, "Licensing Requirements for the Independent Storage of Spent Nuclear Fuel, High-Level Radioactive Waste, Reactor-Related Greater Than Class C Waste," contain licensing requirements to maintain spent fuel integrity. The Commission, in issuing its Waste Confidence Decision in 1990, found that spent fuel can be stored safely in spent fuel pools or in onsite independent spent fuel storage installations without significant environmental impacts for at least 30 years beyond the plant's licensed life (which may include the term of a renewed license).

6.3.10 Reactor Safety Research Program

The NRC conducts reactor safety research to support its mission of ensuring that its licensees safely design, construct, and operate nuclear reactor facilities. The agency carries out this research program to identify, evaluate, and resolve safety issues; to ensure that an independent technical basis exists to review licensee submittals; to evaluate operating experience and results of risk assessments for safety implications; and to support the development and use of risk-informed regulatory approaches. In conducting the Reactor Safety Research Program, the NRC anticipates challenges posed by the introduction of new technologies. The NRC continues to seek out opportunities to leverage its resources through both domestic and international cooperative programs and to provide enhanced opportunities for stakeholder involvement and feedback on its research program.

The NRC conducts pre-application reviews for advanced non-light water reactor designs under the safety research program. In the pre-application phase, NRC interacts with prospective design certification applicants to address topics that would benefit both the applicant and the staff in preparing for a design certification application. The Commission's Policy Statement on Advanced Reactors (SECY-93-087, "Policy, Technical, and Licensing Issues Pertaining to Evolutionary and Advanced Light-Water Reactor Designs,") encourages early interactions on such advanced designs so as to facilitate the resolution of safety issues early in the design process. In addition, the agency will conduct research to address technical issues that it anticipates will arise during its review of advanced reactor designs.

6.3.11 Special Programs for Public Participation

The NRC values public participation in its regulatory processes. To this end, the NRC provides the diverse body of stakeholders (general public; Congress; other Federal, State, and local governments; Indian Tribes; industry; technical societies; the international community; and citizen groups) clear and accurate information about its role and opportunities to participate in the agency's regulatory programs. Numerous NRC programs and processes provide the public with accessibility to NRC staff and resources, seek to make communication with stakeholders more clear, accurate, reliable, objective, and timely, and help to ensure that the reporting of the performance of nuclear power plants performance is open and objective. The agency has developed Web pages to disseminate timely, accurate information regarding issues of interest to the public or events at nuclear facilities. The NRC elicits public involvement early in the regulatory process so as to address any safety concerns in a timely manner. In addition to the formal petition and hearing processes integrated into the licensing program, the agency also uses feedback forms at public meetings to obtain public input.

Fostering an environment in which safety issues can be openly identified without fear of retribution is of paramount interest to the NRC. The agency has established tools for the public, industry, and NRC employees to use to raise safety concerns, including the petition process under 10 CFR 2.206, "Requests for Actions under this Subpart," the safety-conscious work environment policy, and the allegation program.

The NRC petition process regulations in 10 CFR 2.206 allow any member of the public to raise potential health and safety concerns and ask the agency to take specific enforcement actions against a licensee. If warranted, the NRC can modify, suspend, or revoke a license, or take other appropriate enforcement action, to resolve a problem identified in the petition. Recent changes made to the petition process emphasize a timely response to the petitioner and encourage increased, direct involvement of the petitioner (in addition to involvement of the licensee) by allowing the petitioner to personally address the petition review board and comment on the agency's decision.

Any member of the public may petition the NRC to develop, change, or rescind a rule under 10 CFR 2.802. Upon receiving the petition, NRC publishes it in the Federal Register for public comment. The NRC staff will evaluate the petition and any comments received and may either grant or deny the petition or, in some instances, may partially grant/ deny the petition. If granting the petition, the NRC will publish a proposed rule that would address the concern included in the petition. This action would be followed by the publication of a final rule. If denying a petition, NRC publishes a notice of denial in the *Federal Register*. This notice of denial will address any public comments received and the reason for denying the petition.

NRC encourages workers in the nuclear industry to take their concerns directly to their employers and is particularly vigilant about a safety-conscious work environment that encourages such reporting. It is NRC's expectation that licensees establish and maintain a work environment where employees do not fear retribution by a licensee for raising concerns about safety or regulatory issues. Additionally, workers and members of the public may bring their concerns relating to safety or regulatory issues directly to the NRC. The agency established a toll-free safety hotline for reporting such concerns, and NRC management, staff, and inspectors, including the resident inspectors at plant sites, are trained and available to receive such concerns.

Historically, approximately 500 potential allegations have been reported directly to the NRC allegation program each year by industry workers or members of the public. The NRC developed the allegation program to establish a formal process for evaluating and responding to each issue. The primary purpose of the program is to provide an alternative method for individuals to raise safety or regulatory issues and to have them addressed. About 60 percent of the issues that are reported to the NRC are from licensee employees, employees of contractors to licensees, or former employees of licensees or contractors. Given sufficient information, the staff will evaluate each issue to determine whether it can verify the issue and, if so, the effect of the issue on plant safety. The evaluation either involves an engineering review, inspection, or investigation by NRC staff or an evaluation by the licensee that is assessed by the NRC staff.

Historically, the NRC has been able to substantiate 25 to 30 percent of the allegations received. If the evaluation reveals a violation of regulatory requirements, the agency takes appropriate enforcement action. Additionally, the NRC informs in writing the individual who raised the issue of the results of its evaluation, except in limited instances when sensitive security related matters are discussed.

ARTICLE 7. LEGISLATIVE AND REGULATORY FRAMEWORK

1.   **Each Contracting Party shall establish and maintain a legislative and regulatory framework to govern the safety of nuclear installations.**

2.   **The legislative and regulatory framework shall provide for:**

   (i)   the establishment of applicable national safety requirements and regulations

   (ii)   a system of licensing with regard to nuclear installations and the prohibition of the operation of a nuclear installation without a license

   (iii)   a system of regulatory inspection and assessment of nuclear installations to ascertain compliance with applicable regulations and the terms of licences

   (iv)   the enforcement of applicable regulations and of the terms of licences, including suspension, modification, and revocation

This section explains the legislative and regulatory framework governing the U.S. nuclear industry. It discusses the provisions of that framework for establishing national safety requirements and regulations and systems for licensing, inspection, and enforcement. The framework and provisions have not changed since the previous U.S. National Report was issued. This update includes a revised discussion of 10 CFR Part 52.

7.1 Legislative and Regulatory Framework

The Atomic Energy Act, passed by Congress and signed by the President, provided the framework for all subsequent regulation of nuclear installations. However, as is generally the case with most laws, this act provided general principles and concepts and left the regulatory body (i.e., the NRC) to address the details through specific regulations.

7.2 Provisions of the Legislative and Regulatory Framework

7.2.1 National Safety Requirements and Regulations

The regulations on nuclear installations in the United States are governed by the Atomic Energy Act and the Energy Reorganization Act. (The Energy Reorganization Act abolished the U.S. Atomic Energy Commission, and created the NRC and the U.S. Energy Research and Development Administration (ERDA). ERDA was subsequently subsumed by DOE. The NRC administers these statutes by licensing commercial nuclear installations in the United States. In addition, several statutes (listed in previous U.S. National Reports) have substantial bearing on the practices and procedures of the Commission.

The NRC must license all commercial nuclear installations in the United States. (Some Government facilities that are operated by or for DOE are exempt from licensing under the

Atomic Energy Act and the Energy Reorganization Act.) The licensing process is similar in many respects for reactors, separation facilities, reprocessing plants, and nuclear waste storage and disposal facilities. The following discussion describes the licensing practices for nuclear power plants.

7.2.2 Licensing of Nuclear Installations

Chapter 10, Section 103, of the Atomic Energy Act grants the NRC authority to issue licenses for nuclear reactor facilities. In addition, Section 103 states that such licenses are subject to such conditions as the NRC may by rule or regulation establish to effectuate the purposes and provisions of the Atomic Energy Act.

Section 189a. of the Atomic Energy Act provides affected parties with hearing rights in proceedings for the granting, suspending, revoking, or amending of a license or construction permit. Hearings, which are used in licensing proceedings for production and utilization facilities (e.g., nuclear power plants), are held under procedural rules stated in 10 CFR Part 2, "Rules of Practice for Domestic Licensing Proceedings and Issuance of Orders," and, in particular, Subpart C, "Rules of General Applicability." The staff participates as a party in most formal hearings and may also participate as a party in less formal hearings. Hearings are usually held before a three-member Atomic Safety and Licensing Board, which is generally composed of one lawyer and two technical members.

Article 18 describes the licensing process in greater detail. Two alternative approaches to licensing exist. The traditional approach, under 10 CFR Part 50, requires two steps. In the first step, the NRC reviews a preliminary application and decides whether to grant a construction permit. In the second step, the agency reviews the final application and grants an operating license. The NRC licensed all current operating plants in the United States according to this process.

In 1989, the Commission established an alternative licensing system, published in 10 CFR Part 52, which provides for certified standard designs and combined licenses that resolve design issues before construction, and early site permits that resolve most siting issues years before construction. The basic concept underlying 10 CFR Part 52 is that nuclear reactor designs can be approved through generic rulemaking. Once the designs are approved, an applicant can reference them in applications for permission to build and operate nuclear power plants without the necessity to relitigate, in individual hearings, the issues resolved in the rulemaking. Moreover, the NRC will determine and approve before construction the criteria for evaluating whether the plant had been built as specified. Thus, the plant could begin operation without a second hearing, provided that it satisfied the acceptance criteria. To the extent possible, issues would be litigated before construction, not once construction is nearly complete, when the consequences of delay are much greater. In adopting 10 CFR Part 52, the Commission used the latitude allowed by law to streamline licensing.

Recently, NRC amended Part 52 to improve the effectiveness of its processes for licensing future nuclear power plants. The amendments clarify the overall regulatory relationship between Part 50 and Part 52, reorganize Part 52, and reconcile differences in wording in other parts of the regulations to provide consistent terminology throughout all of the regulations affecting part 52. The amendments also added new sections on written communications, employee protection, completeness and accuracy of information, exemptions, combining licenses, and jurisdictional limits.

7.2.3 Inspection and Assessment

Under the Atomic Energy Act, the NRC has the authority to inspect nuclear power plants in its role of protecting public health and safety and the common defense and security. The staff inspects power reactors under construction, in test conditions, and in operation to ascertain compliance with regulations and license conditions. Through its inspection program, the NRC assesses whether activities are properly conducted and equipment is properly maintained to ensure safe operations. The agency integrates inspection results into its overall evaluation of licensee performance, as discussed in Article 6. If a safety problem exists, or there is a failure to comply with requirements, the licensee must take prompt corrective action. If necessary, the NRC may take enforcement action.

7.2.4 Enforcement

The NRC's enforcement jurisdiction is drawn from the Atomic Energy Act and the Energy Reorganization Act.

Section 161 of the Atomic Energy Act authorizes the NRC to conduct inspections and investigations and to issue orders as may be necessary or desirable to promote the common defense and security, protect health, or minimize danger to life or property. Section 186 authorizes the NRC to revoke licenses under certain circumstances (e.g., for material false statements, for a change in conditions that would have warranted NRC refusal to grant a license on an original application, for a licensee's failure to build or operate a facility in accordance with the terms of the permit or license, and for violation of an NRC regulation). Section 234 authorizes the NRC to impose monetary civil penalties not to exceed $100,000 per violation per day; however, that amount is adjusted every 4 years by the Federal Civil Penalties Inflation Adjustment Act and is currently $130,000. In addition to the provisions mentioned in Section 234, Sections 84 and 147 authorize the imposition of civil penalties for violations of regulations implementing those provisions. Section 232 authorizes the NRC to seek injunctive or other equitable relief for violation of regulatory requirements.

Section 206 of the Energy Reorganization Act authorizes the NRC to impose civil penalties on licensees for knowing and conscious failures to provide the agency with certain safety information.

Chapter 18 of the Atomic Energy Act provides for varying levels of criminal penalties (i.e., monetary fines and imprisonment) for willful violations of the act or the regulations or orders issued under Sections 65, 161b, 161i, or 161o of the act.

Section 223 allows criminal penalties to be imposed on certain individuals who are employed by firms constructing or supplying basic components of any utilization facility if the individual knowingly and willfully violates NRC requirements in a manner that could significantly impair a basic component. Section 235 allows criminal penalties to be imposed on persons who interfere with nuclear inspectors. Section 236 allows criminal penalties to be imposed on persons who attempt to or cause sabotage at a nuclear facility or to nuclear fuel. The agency refers alleged or suspected instances of criminal violations of the Atomic Energy Act to the U.S. Department of Justice for appropriate action.

Subpart B, "Procedure for Imposing Requirements by Order, or for Modification, Suspension, or Revocation of a License, or for Imposing Civil Penalties," of 10 CFR Part 2 specifies the procedures that the NRC uses in exercising its enforcement authority. The scope of Subpart B includes the procedures described below:

(1)   10 CFR 2.201, "Notice of Violation," outlines the procedure for issuing notices of violations.

(2)   10 CFR 2.202, "Orders," explains the procedure for issuing orders. In accordance with this section, the NRC may decide to issue an order to institute a proceeding to modify, suspend, or revoke a license or to take other action against a licensee or other person subject to the NRC's jurisdiction. The licensee or any other person adversely affected by the order may request a hearing. The NRC is authorized to make orders immediately effective if required to protect public health, safety, or interest, or if the violation is willful.

(3)   10 CFR 2.204, "Demand for Information," specifies the procedure for issuing a demand for information to a licensee or other person subject to the Commission's jurisdiction to determine whether an order should be issued or other enforcement action should be taken. The demand does not provide hearing rights because only information is being sought. A licensee must answer a demand. An unlicensed person may answer a demand either by providing the requested information or by explaining why the demand should not have been issued.

(4)   10 CFR 2.205, "Civil Penalties," describes the procedure for assessing civil penalties. The NRC initiates the civil penalty process by issuing a notice of violation and proposed imposition of a civil penalty. The agency provides the person charged with an opportunity to contest in writing the proposed imposition of a civil penalty. After evaluating the response, the NRC may mitigate, remit, or impose the civil penalty. If the agency imposes a civil penalty, it provides an opportunity for a hearing. If a civil penalty is not paid following a hearing, or if a hearing is not requested, the agency may refer the matter to the U.S. Department of Justice to institute a civil action in Federal District Court to collect the penalty.

ARTICLE 8. REGULATORY BODY

1.   **Each Contracting Party shall establish or designate a regulatory body entrusted with the implementation of the legislative and regulatory framework referred to in Article 7, and provided with adequate authority, competence, and financial and human resources to fulfill its assigned responsibilities.**

2.   **Each Contracting Party shall take the appropriate steps to ensure an effective separation between the functions of the regulatory body and those of any other body or organization concerned with the promotion or utilization of nuclear energy.**

This section explains the establishment of the U.S. regulatory body (i.e., the NRC). It also explains how the functions of the NRC are separate from those of bodies responsible for promoting and using nuclear energy (e.g., DOE). This update was reorganized. It reports on the establishment of the NRC's Office of New Reactors and Office of Federal and State Materials and Environmental Management Programs; organizational changes; current activities; budget and workforce planning; and the IRRS assessment.

8.1 The Regulatory Body

This section explains the NRC's mandate, authority and responsibilities, structure, international responsibilities and activities, financial and human resources, position in the governmental structure, and report of the IRRS self-assessment team.

8.1.1 Mandate

As discussed in Article 7, Congress created the NRC as an independent regulatory agency in January 1975, with the passage of the Energy Reorganization Act. In giving the NRC an exclusively regulatory mandate, the statute reflected (in part) a Congressional judgment that the expanding commercial nuclear power industry (which was expected to continue to grow) warranted the full-time attention of an exclusively regulatory agency. In creating the NRC, Congress also addressed a developing public concern that regulatory responsibilities were overshadowed by the promotion of nuclear power at the Atomic Energy Commission.

8.1.2 Authority and Responsibilities

8.1.2.1 Scope of Authority

The NRC's mission is to ensure that the civilian uses of nuclear energy and materials in the United States are conducted with proper regard for public health and safety, national security, environmental concerns, and, in the case of the initial licensing of nuclear power plants, the antitrust laws. The Atomic Energy Act provides the charter for these regulatory responsibilities through which Congress created a national policy of developing the peaceful uses of atomic energy. Congress has amended the statute over the years to address developing technology and changing perceptions of regulatory needs. For example, antitrust reviews were added in 1970, the same year that the National Environmental Policy Act of 1969, as amended, imposed broad new responsibilities on Federal agencies. Other more specialized statutes prescribe the

NRC's duties with regard to high-level radioactive waste, low-level radioactive waste, mill tailings, environmental reviews, nonproliferation, antiterrorism, and import/export.

The NRC's licensing authority extends to other government organizations (such as the Tennessee Valley Authority (TVA), which operates nuclear power plants) and the military's use of radiopharmaceuticals in its hospitals. The NRC's responsibilities include both safety and safeguards through which the agency ensures the security of commercial nuclear facilities and materials against radiological sabotage and thefts.

In addition, the NRC is authorized to relinquish its authority, in certain cases, to States (i.e., of the United States) who enter into agreements with the NRC (known as Agreement States).

Section 8.2 of this report provides specific information about the scope of the agency's authority over DOE nuclear installations.

8.1.2.2 The NRC as an Independent Regulatory Agency

The Commission's status as an independent regulatory agency within the Executive Branch of the Federal Government means that its regulatory decisions cannot ordinarily be directed by the President. (By law, however, the U.S. Office of Management and Budget reviews the proposed NRC budget.) Likewise, Congress cannot override the Commission's decisions, except by duly enacted legislation.

The independence of the NRC's decisionmaking process implies a matching responsibility on the part of the Commissioners and their personal staffs to keep the decisionmaking process free from improper outside influence. This is especially important in the case of adjudications. When the Commissioners take part in adjudications, they ordinarily act in the role of appellate judges (reviewing the decisions of lower judges) and, in general, are bound by the same kinds of strictures that apply to judges in Federal courts.

8.1.3 Structure of the Regulatory Body

This section explains the structure of the NRC. It covers the Commission, component offices and their responsibilities, and advisory committees and their functions. It also explains recent changes in NRC organization.

8.1.3.1 The Commission

The NRC is headed by a five-member Commission. The President designates one member to serve as Chairman and official spokesperson. The Commission as a whole formulates policies and regulations governing nuclear reactor and materials safety, issues orders to licensees, and adjudicates legal matters brought before it. The Executive Director for Operations carries out the policies and decisions of the Commission and directs the activities of the program offices.

8.1.3.2 Component Offices of the Commission

Since the previous U.S. National Report was issued, the NRC has reorganized and established two new offices—the Office of New Reactors and the Office of Federal and State Materials and

Environmental Management Programs. These offices, as well as other NRC offices, are briefly described in the following paragraphs.

Office of the Executive Director for Operations. The Executive Director for Operations is the chief operational and administrative officer of the Commission and is authorized and directed to discharge such licensing, regulatory, and administrative functions and to take such actions as necessary for day-to-day operations of the agency. The Executive Director supervises and coordinates policy development and operational activities of NRC program and regional offices and implements Commission policy directives pertaining to these offices.

Office of the Chief Financial Officer. The Office of the Chief Financial Officer is responsible for the NRC's planning and budgeting and performance management process and for all NRC financial management activities.

Office of the General Counsel. The Office of the General Counsel directs matters of law and legal policy, providing opinions, advice, and assistance to the agency on all of its activities.

Office of the Inspector General. The Inspector General provides leadership and policy direction in conducting audits and investigations to promote economy, efficiency, and effectiveness within the NRC and to prevent and detect fraud, waste, abuse, and mismanagement in agency programs and operations.

Office of International Programs. This office provides assistance and recommendations to the Chairman, the Commission, and the NRC staff on international issues. It coordinates the NRC's international activities; plans and implements policies in the international arena; and establishes and maintains working relationships with individual countries and international nuclear organizations as well as other involved U.S. Government agencies.

Office of Public Affairs. The Office of Public Affairs directs the agency's public affairs program, advising agency officials and developing key strategies that help increase public confidence in NRC policies and activities.

Office of Congressional Affairs. The Office of Congressional Affairs is the primary point of contact for all communications between the NRC and Congress. This office monitors legislative proposals, bills, and hearings; informs NRC of the views of Congress on NRC policies, plans, and activities; provides timely responses to congressional requests for information; and provides the information necessary to keep appropriate Members of Congress and congressional staff fully and currently informed of NRC actions.

The Office of Commission Appellate Adjudication. This office provides the Commission with an analysis of any adjudicatory matter requiring a Commission decision and drafts necessary decisions pursuant to the Commission's guidance after presentation of options.

Office of the Secretary of the Commission. This office provides executive management services to support the Commission and to carry out Commission decisions. It assists with the planning, scheduling, and conduct of Commission business; maintains historical paper files of official Commission records; administers the NRC Historical Program; and maintains the Commission's official adjudicatory and rulemaking dockets.

8.1.3.3 Offices of the Executive Director for Operations

The offices reporting to the Executive Director for Operations ensure that the commercial use of nuclear materials in the United States is safely conducted.

Office of Nuclear Reactor Regulation. The Office of Nuclear Reactor Regulation (NRR) is responsible for accomplishing key components of the NRC's nuclear reactor safety mission. To do so, NRR conducts a broad range of regulatory activities in the four primary program areas of rulemaking, licensing, oversight, and incident response for commercial nuclear power reactors and test and research reactors to protect the public health, safety, and the environment.

Office of New Reactors. This office is responsible for accomplishing key components of the NRC's nuclear reactor safety mission for new reactor facilities licensed in accordance with 10 CFR Part 52. As such, NRO is responsible for regulatory activities in the primary program areas of siting, licensing and oversight for new commercial nuclear power reactors.

Office of Nuclear Material Safety and Safeguards. The Office of Nuclear Material Safety and Safeguards is responsible for regulating activities that provide for the safe and secure production of nuclear fuel used in commercial nuclear reactors; the safe storage, transportation and disposal of high-level radioactive waste and spent nuclear fuel; and the transportation of radioactive materials regulated under the Atomic Energy Act.

Office of Nuclear Security and Incident Response. This office develops overall agency policy and provides management direction for evaluating and assessing technical issues involving security and emergency preparedness at nuclear facilities.

Office of Nuclear Regulatory Research. This office plans, recommends, and conducts research programs to identify, lead, and sponsor reviews that support the resolution of ongoing and future safety issues.

Regional Offices. The four regional offices conduct inspection, enforcement, and emergency response programs for licensees within their borders.

Office of Enforcement. This office oversees, manages, and directs the development and implementation of policies and programs for enforcing NRC requirements. It oversees the agency's allegations management programs and the allegations review process. The office is responsible for external safety culture policy matters, the agency's Alternative Dispute Resolution program, the agency's internal Differing Professional Opinions Program, and its new internal nonconcurrence process.

Office of Investigations. This office develops policy, procedures, and quality control standards for investigations of licensees and applicants, as well as their contractors or vendors, including the investigation of all allegations of wrongdoing by non-NRC employees and contractors.

Office of Federal and State Materials and Environmental Management Programs. This office is responsible for effective communications and working relationships between NRC and other governmental entities and administers the Agreement State Program (through which States have signed formal agreements with the NRC, to assume regulatory responsibility over certain byproduct, source, and small quantities of special nuclear material). It also develops and implements rules and guidance for the safe and secure use of source, byproduct and special nuclear material in industrial, medical, academic, and commercial activities, and at decommissioning, uranium recovery, and low-level waste sites.

This office was formerly the Office of State and Tribal Programs, but was reorganized to merge with part of the Office of Nuclear Material Safety and Safeguards. This reorganization enhanced the integration of the National Materials Program, a broad collective framework within which both NRC and the Agreement States carry out their respective radiation safety regulatory programs. The reorganization recognizes the increasing number of Agreement States, the value of their experience in administering the National Materials Program, and the importance of coordination between the NRC, the States, and other stakeholders.

Office of Information Services. This office plans, directs, and oversees the delivery of centralized information technology infrastructure, applications, and information management services, in addition to the development and implementation of information technology and management plans, architecture, and policies to support the mission, goals, and priorities of the agency.

Support Offices. Supporting the Executive Director of Operations are the Offices of Administration, Human Resources, and Small Business and Civil Rights.

8.1.3.4 Advisory Committees

The three principal advisory committees for NRC programs are the Advisory Committee on Reactor Safeguards, the Advisory Committee on Nuclear Waste, and the Advisory Committee on Medical Uses of Isotopes. In addition, the NRC has established an ad hoc Licensing Support Network Advisory Panel. The Advisory Committee on Reactor Safeguards is the only committee relevant to this report. This committee reviews and reports on safety studies and reactor facility license and license renewal applications; advises the Commission on the hazards of proposed and existing reactor facilities and the adequacy of proposed reactor safety standards; initiates reviews of specific generic matters or nuclear facility safety-related items; and reviews the NRC's research activities.

8.1.4 International Responsibilities and Activities

The NRC conducts international activities related to statutory mandates, international treaties, conventions, codes, international organizations, bilateral relations, and research.

U.S. law or international treaties and conventions mandate several NRC international activities; other activities are discretionary. In particular, the NRC is statutorily mandated to serve as the U.S. licensing authority for exports and imports of nuclear materials and equipment.

The NRC supports U.S. foreign policies in the safe and secure use of nuclear materials and in guarding against the spread of nuclear weapons. The agency actively participates in developing and implementing a variety of legally binding treaties and conventions which create an international framework for the peaceful uses of nuclear energy. The NRC provides technical and legal advice and assistance to international organizations and foreign countries as they work to develop effective regulatory organizations and rigorous safety standards. Some activities are carried out within the programs of the IAEA, the Nuclear Energy Agency of the Organization for

Economic Cooperation and Development (NEA), or other international organizations. Others are conducted directly with counterpart agencies in other countries under cooperation agreements.

Export-Import. The NRC's key international responsibility is licensing the export and import of nuclear materials and equipment, such as low-enriched uranium fuel for nuclear power plants, high-enriched uranium for research and test reactors, nuclear reactors themselves, certain nuclear reactor components (such as pumps and valves), and radioisotopes used in industrial, medical, agricultural and scientific fields. The NRC ensures that such exports and imports are consistent with the goals of the safe and peaceful use of these materials and equipment, limiting the proliferation of nuclear weapons, and promoting the Nation's common defense and security. The Atomic Energy Act, the Nuclear Non-Proliferation Act of 1978, and 10 CFR Part 110, "Export and Import of Nuclear Equipment and Material," detail the standards and procedures for issuing export and import licenses.

International Treaties. Treaties that legally bind the NRC and the U.S. Government's peaceful uses of nuclear energy include the 1978 Nuclear Non-Proliferation Treaty, the 1980 Convention on Physical Protection of Nuclear Material, the 1994 Convention on Nuclear Safety, the 1986 Convention on Early Notification of a Nuclear Accident, the 1986 Convention on Assistance in Case of a Nuclear Accident or Radiological Emergency, and the 1997 Joint Convention on the Safety of Spent Fuel Management and on the Safety of Radioactive Waste Management. NRC staff regularly participate in international meetings related to these conventions and have held a variety of Convention leadership positions. In its bilateral work with regulatory counterparts worldwide, the NRC works to exchange experience and good practices in order to further the goals of these international instruments.

International Organizations and Associations. The NRC actively participates in the full scope of programs of the two major international nuclear organizations, the IAEA and the NEA. For example, since 1996, the United States has or is planning to participate in more than 30 Operational Safety Assessment Review Team (OSART) missions. Some experts on these teams come from the NRC, while others come from industry. The NRC coordinates closely with INPO in this process. Since 1999, the NRC has participated in approximately 15 IRRT or IRRS missions and intends to participate in at least four additional missions before the end of 2007. The NRC is currently coordinating with the IAEA and industry in planning an OSART mission to Arkansas Nuclear One Nuclear Power Plant in 2008 and intends to continue to plan for an OSART mission in the United States every 3 years. The NRC also participates in the Commission on Safety Standards and safety standards committees. Other IAEA activities include the extra-budgetary program for long-term operation.

The NRC holds leadership roles in the four IAEA Safety Standards Committees and the Commission on Safety Standards. These activities, together with regular NRC staff participation in IAEA meetings to draft and revise safety and security guidance and coordination with other U.S. Government agencies, enable the NRC to use its broad regulatory experience to contribute to the safe and secure use of nuclear and radioactive materials in IAEA Member States. The NRC also participates in the NEA Steering Committee, and holds leadership positions in the Committee on the Safety of Nuclear Installations, the Committee on Nuclear Regulatory Activities, the Committee on Radiation Protection and Public Health, and the Radioactive Waste Management Committee. The NRC is also represented on many of the NEA committee-chartered working groups. These activities provide diverse fora for nuclear regulators and research organizations to share information and work together to leverage resources for mutual benefit.

The NRC continues to participate in the Multinational Design Evaluation Program (MDEP) with the goal of leveraging the experience of international counterparts in the review of new reactor designs. Currently, the NRC is participating in two parallel MDEP activities. In Stage 1 of the MDEP, the NRC is cooperating with the regulatory authorities in Finland and France on the design reviews of the AREVA EPR. This reactor is currently being constructed in Finland, has been approved for construction in France, and an application for design certification will be submitted to the NRC in 2008. As part of Stage 2 of the MDEP, the NRC is participating with nine other countries in a program to assess ways to enhance cooperation through the better understanding and possible convergence of the country specific regulations, codes, and standards associated with the design reviews of new reactors.

In addition to staff participation in more than 100 IAEA and Nuclear Energy Agency meetings each year, the NRC Chairman routinely participates in the IAEA General Conference and biannual meetings of the International Nuclear Regulators Association. Members of the Commission also travel to international conferences around the world to deliver keynote remarks, participate in panel discussions, and otherwise share insight on a variety of topics to diverse international audiences.

Bilateral Relations. The NRC works closely with nuclear safety agencies in more than 35 countries. The NRC and its foreign counterparts routinely exchange operational safety data and other regulatory information. Subject to outside funding, the NRC provides safety, security, emergency preparedness and safeguards advice, training, and other assistance to countries that seek U.S. help to improve their regulatory programs.

The NRC's information exchange arrangements serve as communication channels with foreign regulatory authorities, establishing a framework for the agency to gain access to non-U.S. safety information which can (1) alert the U.S. Government and industry to potential safety problems, (2) help identify possible accident precursors, and (3) provide accident and incident analyses, including lessons learned, which could be directly applicable to the safety of U.S. nuclear power plants and other facilities. The arrangements also serve as vehicles for the assistance the NRC provides to countries to establish and improve their regulatory capabilities and infrastructure. Thus, the arrangements facilitate NRC's strategic goal to support U.S. interests in the safe and secure use of nuclear materials and in nuclear nonproliferation. There are currently 38 active bilateral arrangements between the NRC and its foreign regulatory counterparts. These arrangements allow the staff to conduct regular bilateral exchanges on a variety of levels. NRC also has bilateral interactions with countries with which there is no active arrangement, though the absence of an arrangement limits the type of information that can be exchanged. The NRC Chairman typically meets with at least 20 foreign counterparts at the IAEA's annual General Conference. In addition, members of the Commission travel abroad to hold bilateral meetings with their regulatory counterparts, tour nuclear power plants and other facilities, and exchange information and good practices. Often, these visits result in increased communication between the NRC and its counterparts, providing opportunities for enhanced information exchange based on first-hand knowledge of various programs.

International Assistance Programs. In the early 1990s, the NRC began offering assistance to nuclear regulatory programs in several former Soviet states. Efforts were initially focused on those countries in which Soviet-designed reactors were operated. Following the September 11,

2001 terrorist attacks, assistance efforts were expanded to specifically include assisting countries in their efforts to improve regulatory oversight of radioactive sources. In addition, the NRC is assisting the Government of Iraq in its efforts to develop a sound regulatory structure, including the provision of assistance in developing the law and regulations that will be the legal framework for the project of decommissioning former facilities that used radioactive materials in Iraq.

Research Programs. The NRC conducts confirmatory regulatory research through more than 90 multilateral agreements in partnership with nuclear safety agencies and institutes in more than 20 countries. This research supports regulatory decisions on emerging technologies, aging equipment and facilities, and various other safety issues. The NRC and other nuclear regulatory and safety organizations carry out cooperative research projects to achieve mutual research needs with greater efficiency. This research is currently conducted under the following programs: Cooperative Severe Accident Research Program; Code Applications and Maintenance Program; and Steam Generator Tube Integrity Program.

8.1.5 Financial and Human Resources

8.1.5.1 Financial Resources

As of September 30, 2006, and 2005, the financial condition of the NRC was sound with respect to having sufficient funds to meet program needs and adequate control of these funds in place to ensure obligations did not exceed budget authority. The sum of all funds available to obligate for FY 2006 was $809.0 million, which is a $86.1 million increase over the FY 2005 amount of $722.9 million.

The NRC FY 2008 budget will be financed with $765.1 million from user fees, $114.3 million from the General Fund, and $37.3 million from the Nuclear Waste Fund.

8.1.5.2 Human Resources

The NRC was ranked as the best place to work in the federal government among large federal agencies in 2007. This ranking, along with new recruiting authority provided by Congress, should assist in the agency's hiring efforts to maintain an innovative and effective workforce. The NRC is recruiting about 400 employees each year for the next few years because of the expected arrival of close to two dozen applications for new reactor licenses beginning in the fall of 2007. The NRC is working to aggressively meet the challenge of recruiting, hiring, training and integrating new employees with the necessary infrastructure to support its activities, as well as retaining staff, as discussed below.

Recruitment and Hiring Process. To meet the demands of hiring new employees and to address the need for more experienced individuals, the NRC identified the need to expand its recruitment activities and streamline the hiring process. As a long standing practice, NRC actively recruits for its Nuclear Safety Professional Development Program at targeted universities with a history of graduating technically strong, diverse candidates. It has expanded its recruitment activities at professional society conferences and career fairs, and is expanding advertising in trade journals and on Web sites to attract professionals in specialized technical disciplines and in local newspapers around the country where technical engineers and scientists may be interested in re-locating due to job cutbacks in their areas.

The Agency has streamlined recruitment, relocation, and retention incentives to allow offices to extend job and incentive offers to outside applicants. For example, NRC has revised the generic open vacancy announcement for mid-career engineers and scientists to provide additional flexibilities to offer relocation and recruitment incentives. The NRC is also creating the policy of offering referral awards. Other innovations, such as student loan repayments, waivers of dual compensation limitations, partnerships with colleges and universities, and the Cooperative Education Program, have an equally positive impact on the Agency's efforts to recruit and retain staff with critical skills.

Training and Knowledge Management (KM)/transfer. About 40% of staff have been with the NRC for less than 4 years. The rapid integration and training of such a large number of new employees into the Agency is a significant challenge but is essential for the NRC's and the employees' future success and productivity. To address this challenge, the NRC is expanding the use of existing training tools, including mentoring, on-the-job-training, formal classroom and on-line training, and self-study activities. Senior staff are being asked to help train newer staff in addition to their regular workload and to help new employees assimilate into the NRC culture. A major challenge is the multi-generational population now working together, each with different ways of approaching work.

NRR developed an integrated approach to training to provide new employees with consistent information from branch to branch and division to division. NRR developed a qualification program that consists of three parts: general requirements, position specific requirements, and an oral confirmation board. The training is focused on developing engineers and scientists into regulators. To ensure NRR's administrative staff continues to excel, an administrative support training program was developed.

To help new employees adjust to the NRC, an "NRR New Employee Orientation and Training Guide" has been developed and is being implemented. The guide includes training courses, reading assignments and self-study activities. Additionally, new employees will be assigned a "docent" that will be a peer, typically from their branch, to assist in adjusting. Additionally, position-specific training is expected to accompany the generic training in the "NRR New Employee Orientation and Training Guide." The NRC is continuing to develop its qualification plans and other position-specific training, for example, for project engineers/managers. It is also identifying course needs at its Technical Training Center and Professional Development Center.

For succession planning and knowledge management in critical skills and knowledge areas, supervisors and managers have been provided a new tool in the NRC's Strategic Workforce Planning system to help identify skill gaps. Using the skill categories and the needs assessment existing in the system, supervisors can view a staff and critical skill matrix. This matrix displays the branch employee's level of expertise in each of the most critical skills identified by the branch chief. By identifying potential skill gaps, supervisors can make more informed decisions when assigning work, and can identify skill areas for individual employee development.

Workforce Planning and Deployment With a renewed emphasis on hiring to meet the expected increase in new reactor work, NRR realigned to emphasize the area of new reactors, and the Office of Nuclear Material Safety and Safeguards realigned to enhance cooperation with States and implement a holistic approach to fuel issues including transportation, storage, and disposal. The changes in these two offices were made easier by the NRC's strategic workforce planning tool which allowed for a smoother planning process to improve workforce deployment, maintain technical capacity, and make informed decisions on human capital strategies for recruitment, development, and retention.

Leadership and KM. The NRC continues to offer and expand its leadership competency development programs, such as executive leadership seminars, the Senior Executive Service Candidate Development Program, leadership training for new supervisors and team leaders, and the Leadership Potential Program.

KM is a part of the strategic management of human capital, along with strategic workforce planning, recruitment, and training and development. As part of this effort, the NRC is in the process of coordinating its efforts to implement KM strategies, including developing a KM Web site.

In addition, the NRC is developing an Agencywide KM plan that will serve as a framework to integrate new and existing approaches that generate, capture, and transfer knowledge and information relevant to the NRC's mission. The following are some of the near-term and long-term strategies for this plan:

• capture relevant critical knowledge of departing personnel

• recapture departed knowledge where possible

• communicate leadership's expectation for a knowledge-sharing culture

• formalize KM values and principles

• incorporate KM within process work flows

Following are some of the current Knowledge Management/Knowledge Transfer Activities.

- Branch Chief/Team Leader Seminars. As the role and expectations of branch chiefs evolved from senior technical experts to managers, it is essential that they have the information they need to succeed in their positions. As a community of practice, the branch chiefs/team leaders meet monthly, and are given presentations by agency experts in topics such as performance management, budget, and communications.

- Branch/Team Meetings. To ensure staff in each branch or team are kept up-to-date in areas under their purview, branch chiefs and team leaders hold regularly scheduled staff meetings. During some of these meetings, senior staff members are asked to give presentations to staff regarding an area in which they are considered experts, or passing their knowledge of past events on to newer staff. Some branch chiefs also have their more junior staff give presentations. This is also a way for junior and senior staff to interact, as the junior staff member may be required to interview more senior staff members to glean information for their presentations.

- Video Interviews. NRC conducted a pilot to capture knowledge from retiring senior staff using video interviews. One video captured knowledge regarding steam generators; another was entitled "Nuclear Knowledge for the Next Generation." The interviews included questions about licensing issues, recruiting and mentoring new hires, leadership, operations center experience, and reactor licensing performance metrics.

- Web Sites NRR has a very active web site devoted to KM. Entitled "Sharing Expert Experience and Knowledge" (SEEK), this site contains information such as the Inspector Best Practices Booklet and Inspector Newletters, supervisor and team leader seminars, new employee orientation and training guide, key reference materials for reviews, qualification plans, strategic workforce planning, KM, and other communities of practice.

- Communities of Practice. NRR is piloting an electronic community of practice Demonstrations have been given to a technical, regulatory and policy consistency working group for consideration. Demonstrations to other groups who may benefit by its use continue to be given.

Retaining Staff. Ability to retain recent recruits as opportunities arise in the industry is a challenge, because people are no longer committed to working in one place for their entire career. However, challenging work alone does not ensure retention. The NRC believes that the benefits of the workplace will more than likely help in retaining staff. To this end, NRC continues to try to create an environment where people feel valued and challenged. For example, NRC has made work hours flexible and the ability to work out in the on-site fitness center to help staff create a balance in their life. It has created an organization that the staff feels a part of; one that lives up to its values and has an open environment where communication flows in all directions, and challenging the status quo is encouraged.

8.1.6 Position of the NRC in the Governmental Structure

This section explains the relationship of the NRC to the Executive Branch, the States, and Congress.

8.1.6.1 Executive Branch

The components of the Executive Branch with which the NRC has the most frequent contact and interaction are the White House, Office of Management and Budget, Department of State, DOE, EPA, FEMA, Department of Labor, Department of Transportation, and Department of Justice. Section 8.2 of this report discusses the NRC's relationship to DOE. The agency's relationships with other components of the Federal Government are described below.

The White House. As noted above, as an independent regulatory agency, the White House cannot directly set NRC policy. It can, however, influence NRC policy by (1) appointing Commissioners and Chairmen in whose outlook and judgment it has confidence and (2) making its views known on nonadjudicatory matters. In certain areas, such as national security policy, the Commission has declared its intent to give great weight to the views of the Executive Branch. In informal policy matters, such as rulemaking, White House and Executive Branch officials may properly try to influence NRC decisions, either publicly or privately. Ultimately, however, the NRC must make the decision and accept responsibility for it.

Office of Management and Budget. The NRC submits its annual budget requests, including proposed personnel ceilings, to the Office of Management and Budget for approval.

U.S. Department of State. By law, the NRC must license the export and import of nuclear equipment and material. For significant applications, the Commission requests the U.S. Department of State to provide Executive Branch views on whether the license should be issued.

The NRC also works with the State Department negotiating international agreements in the nuclear field and interacting with the IAEA and other international organizations of the United Nations, as well as the Nuclear Energy Agency of the Organization for Economic Cooperation and Development. In general, the purposes of these interactions are to develop policy on nuclear issues that are under NRC purview and to plan and coordinate programs of nuclear safety and safeguards assistance to other countries.

EPA. The responsibilities of NRC and EPA intersect or overlap in areas in which the EPA issues generally applicable environmental standards for activities that are also subject to NRC licensing. Examples include standards for high-level waste repositories, decommissioning standards, and standards for public and worker protection. The EPA has the ultimate authority to establish generally applicable environmental standards to protect the environment from radioactive material.

FEMA FEMA assists the NRC's licensing process by preparing reviews and evaluations, as well as presenting witnesses to testify at licensing hearings. FEMA also participates with the NRC in observing and evaluating emergency exercises at nuclear plants. FEMA findings are not binding on the NRC, but they are presumed to be valid unless controverted by more persuasive evidence. FEMA is now part of DHS.

Department of Transportation. The NRC and the Department of Transportation share responsibility for the control of radioactive material transport. Department of Transportation regulations cover all aspects of transportation, including packaging, shipper and carrier responsibilities, documentation, and all levels of radioactive material.

U.S. Department of Labor. The NRC monitors discrimination actions related to NRC-licensed activities filed with the Department of Labor under Section 211 of the Energy Reorganization Act and develops enforcement actions when there are properly supported findings of discrimination, either from the NRC's Office of Investigations or from U.S. Department of Labor adjudications.

U.S. Department of Justice. Any NRC litigation almost always requires coordination with the U.S. Department of Justice. Under the Administrative Orders Review Act (commonly called the Hobbs Act), the United States is a party to petitions for review challenging NRC licensing decisions or regulations.

The Office of Investigations, which investigates allegations of wrongdoing by NRC applicants and licensees, as well as by their contractors, normally works with the General Litigation Section of the Criminal Division at the Department's Headquarters and with U.S. Attorneys in the field.

The Office of the Inspector General reports to the Department of Justice whenever it has reasonable grounds to believe that an NRC employee or contractor has violated Federal law. The Inspector General refers cases for review for possible criminal prosecution to the U.S. Attorney's Office for the area in which the potential violation occurred. When the Department of Justice desires support from the Office of the Inspector General for investigations or grand jury work, it makes the request directly to the Inspector General.

8.1.6.2 The States (i.e., of the United States)

At the NRC, the Office of Federal and State Materials and Environmental Management Programs is responsible for establishing and maintaining effective communications and working relationships between the NRC and the States and serves as the primary contact for policy matters, keeping the States informed about NRC activities and keeping the NRC appraised of State activities and views that may affect NRC policies, plans, and activities. Other NRC offices provide major support to implement State relations program policy and guidance, for example, through regional state liaisons and State agreements officers.

As explained above, the Atomic Energy Act confers on NRC preemptive authority over health and safety regulation of nuclear energy and Atomic Energy Act materials. As a result, the general rule is that nuclear power plant safety, like airline safety, is the exclusive province of the Federal Government and cannot be regulated by the States. A State law that attempted to set nuclear safety standards would thus be voided by the courts. However, the courts will not overturn a State law that regulates nuclear energy for purposes other than health and safety, such as economics, unless it conflicts with an NRC requirement. Similarly, the courts will not ordinarily question a State's declared purpose in enacting legislation.

However, the Atomic Energy Act did not entirely exclude States from the regulation of nuclear matters. Section 274 of the act created the Agreement State Program, under which the NRC may relinquish its authority over most nuclear materials to those States willing to assume that authority. The NRC may not relinquish authority over such facilities as reactors, fuel reprocessing and enrichment plants, imports and exports, critical mass quantities of special nuclear material, high-level waste disposal, or certain other excepted areas.

Many States have signed formal agreements with the NRC and have assumed regulatory responsibility over certain byproduct, source, and small quantities of special nuclear material. Agreement States receive no Federal funding to support their regulatory programs. The NRC conducts performance-based reviews of Agreement State programs to ensure that they remain adequate to protect the public health and safety and are compatible with the NRC materials program.

Some States have shown a desire to participate in matters relating to nuclear power plants. In response, the NRC issued a policy statement in February 1989 declaring its intent to cooperate with States in the area of nuclear power plant safety by keeping States informed of matters of interest to them and considering proposals for State officials to participate in NRC inspection activities, pursuant to a memorandum of understanding between the State and NRC. The policy statement makes clear that States must channel their contacts with the NRC through a single State Liaison Officer, appointed by the Governor. States are authorized only to observe and assist in NRC inspections of reactors and they cannot conduct their own independent health and safety inspections.

Through the Intergovernmental Liaison Program, the NRC works in cooperation with Federal, State, and local governments; interstate organizations; and Native American Tribal Governments to maintain effective relations and communications with these organizations and to promote greater awareness and mutual understanding of the policies, activities, and concerns of all parties involved as they relate to radiological safety at NRC-licensed facilities.

8.1.6.3 Congress

The following paragraphs discuss the NRC oversight committees in the Senate and House and subcommittees with jurisdiction over aspects of NRC's activities.

<u>Senate Oversight</u>. In the U.S. Senate, the Committee on the Environment and Public Works has jurisdiction over domestic nuclear regulatory activities. With the committee, the Subcommittee on Clean Air and Nuclear Safety has responsibility for regulation and oversight of the NRC. The Energy and Natural Resources Committee and the Environment and Public Works Committee share jurisdiction over nuclear waste issues.

<u>House Oversight</u>. In the U.S. House of Representatives, the Committee on Energy and Commerce has jurisdiction over domestic nuclear regulatory activities. Within the committee, the Subcommittee on Energy and Air Quality has responsibility for regulation and oversight of the NRC.

<u>Other Relevant Committees</u>. In addition to the committees and subcommittees mentioned above, the House and Senate Appropriations Subcommittees on Energy and Water Development play a key role in approving the Commission's annual budget. Also, a number of committees frequently interface with the NRC concerning international affairs, research, security, and general governmental operations.

8.1.7 Report of the IRRS Self-Assessment Team

At the third review meeting of the Convention, the U.S. committed to perform an IRRT (now called the IRRS) self-assessment.

In the fall of 2006, an inter-office IRRS self-assessment team of about 14 people was constituted. The team consisted of representatives from the Offices of: Nuclear Reactor Regulation, Nuclear Regulatory Research, Federal and State Materials and Environmental Management Programs, Nuclear Materials Safety and Safeguards, Nuclear Security and Incident Response, General Counsel, Chief Financial Officer, Human Resources and International Programs.

The IRRS self-assessment questionnaire was developed in collaboration with the Canadian Nuclear Safety Commission and the IAEA and based on the IRRT questionnaire (IAEA TECDOC-703) and IAEA draft document entitled "Regulatory Performance Self-Assessment Tool." Supplemental questions were added to help assess the regulatory readiness for potential new reactor licensing. The IRRS self-assessment questionnaire consists of over 275 questions.

The team reviewed the results of the IRRS self-assessment responses and developed 12 high-level recommendations. These recommendations are being reviewed by a senior management review board, which will propose appropriate staff treatment in response to them.

8.2 <u>Separation of Functions of the Regulatory Body from Those of Bodies Promoting Nuclear Energy</u>

Although both the NRC and DOE have responsibilities for managing nuclear facilities and materials, they maintain separate, independent functions. The partitioning of the U.S. Atomic

Energy Commission in the mid-1970s provided distinct entities for the U.S. Government's regulatory and promotional responsibilities in nuclear applications.

Specifically, the Energy Reorganization Act redistributed the functions performed by the U.S. Atomic Energy Commission to two new agencies. This act created the NRC to regulate the commercial nuclear power sector and ERDA to promote energy and nuclear power development and to develop defense applications. The NRC was established as an independent authority to regulate the possession and use of nuclear materials as well as the siting, construction, and operation of nuclear facilities. ERDA was established to ensure the development of all energy sources, increase efficiency and reliability of energy resource use, and carry out the other functions, including but not limited to the U.S. Atomic Energy Commission military and production activities and general basic research activities.

The NRC performed its regulatory mission by issuing regulations, licensing commercial nuclear reactor construction and operation, licensing the possession of and/or use of nuclear materials and wastes, safeguarding nuclear materials and facilities from theft and radiological sabotage, inspecting nuclear facilities, and enforcing regulations. The NRC regulates the commercial nuclear fuel cycle materials and facilities. Regarding the regulatory control of commercial spent nuclear fuel and radioactive waste, the NRC is responsible for licensing commercial nuclear waste management facilities, independent spent fuel management facilities, and the DOE-proposed Yucca Mountain site for the disposal of high-level waste and spent fuel.

DOE addresses the U.S. Government's need to unify energy organization and planning. The DOE Organization Act brought a number of the Federal Government's agencies and programs, including ERDA, into a single agency with responsibilities for nuclear energy technology and nuclear weapons programs. Over the past decade, DOE has added new nuclear-related activities directed to environmental cleanup of contaminated sites and surplus facilities and nonproliferation. DOE retains authority under the Atomic Energy Act for regulating its nuclear activities. DOE also retains responsibility for regulating the disposal of its own low-level radioactive waste.

ARTICLE 9. RESPONSIBILITY OF THE LICENSE HOLDER

Each Contracting Party shall ensure that prime responsibility for the safety of a nuclear installation rests with the holder of the relevant license and shall take the appropriate steps to ensure that each such license holder meets its responsibility.

The NRC, through the Atomic Energy Act, ensures that the prime responsibility for the safety of a nuclear installation rests with the licensee. Steps the NRC takes to ensure that each licensee meets its primary responsibility include the licensing process, discussed in Articles 18 and 19, the Reactor Oversight Process, discussed in Article 6, and the Enforcement Program, discussed below. This update revises the debt collection dollar amount, discusses the alternative dispute resolution program and current experience.

9.1 Introduction

The NRC's regulatory programs continue to be based on the premise that the safety of commercial nuclear power reactor operations is the responsibility of NRC licensees. The responsibility of the NRC is regulatory oversight of licensee activities to ensure that safety is maintained. The NRC reviews the safety of a reactor design and the capability of an applicant to design, construct, and operate a facility. If an applicant satisfies the requirements of the *Code of Federal Regulations*, the NRC then issues a license to operate the facility. Such licenses specify the terms and conditions of operation to which a licensee must conform. Failure to conform subjects the licensee to enforcement action, which can include modifying, suspending, or revoking the license. The NRC can also order particular corrective actions or issue civil penalties. The following paragraphs discuss these enforcement mechanisms in greater detail.

9.2 The Licensee's Prime Responsibility for Safety

As discussed in Article 7 of this report, Chapter 10, Section 103 of the Atomic Energy Act grants the NRC authority to issue licenses for nuclear reactor facilities. Moreover, Section 103 states that these licenses are subject to such conditions as the NRC may by rule or regulation establish to effectuate the purposes and provisions of the Atomic Energy Act. Consistent with the act, before issuing a license, the Commission determines that the applicant is (1) equipped and agrees to observe such safety standards to protect health and to minimize danger to life or property as the Commission may by rule establish and (2) agrees to make available to the Commission such technical information and data about activities under such license as the Commission may determine necessary to promote the common defense and security and to protect the health and safety of the public.

Embedded in each license is the explicit responsibility that the license holder comply with the terms and conditions of the license and the applicable Commission rules and regulations. The licensee is ultimately responsible for the safety of its activities and the safeguarding of nuclear facilities and materials used in operation.

When the Commission or licensee determines that the licensee is not complying with the Commission's rules or regulations, action is taken to ensure that the facility is returned to a condition compliant with its license.

9.3 NRC Enforcement Program

As discussed in Article 7, the NRC has enforcement powers. As discussed in Section 6.2.2.5, the enforcement process complements the Reactor Oversight Process. The NRC uses enforcement as a deterrent to emphasize the importance of compliance with regulatory requirements and to encourage prompt identification and prompt, comprehensive correction of violations.

The NRC identifies violations through inspections and investigations. All violations are subject to civil enforcement action and may be subject to criminal prosecution. Unlike the burden of proof standard for criminal actions (beyond a reasonable doubt), the NRC uses the Administrative Procedure Act standard (preponderance of evidence) in enforcement proceedings. After an apparent violation is identified, it is assessed in accordance with the Commission's enforcement policy, found in NUREG-1600, "General Statement of Policy and Procedures for NRC Enforcement Actions," issued July 2000, which is available to NRC licensees and members of the public. The NRC's Office of Enforcement maintains the current policy statement on NRC's public Web site. Because it is a policy statement and not a regulation, the Commission may deviate from it, as appropriate, given the circumstances of a particular case.

The NRC has three primary enforcement sanctions available. Specifically, those sanctions include notices of violation, civil penalties, and orders.[2] A notice of violation identifies a requirement and how it was violated, formalizes a violation pursuant to 10 CFR 2.201, "Notice of Violation," requires corrective action, and normally requires a written response. A civil penalty is a monetary fine issued under authority of Section 234 of the Atomic Energy Act or Section 206 of the Energy Reorganization Act. Section 234 of the Atomic Energy Act provides for penalties of up to $100,000 per violation per day; however, that amount is adjusted every 4 years by the Federal Civil Penalties Inflation Adjustment Act of 1990, as amended by the Debt Collection Improvement Act of 1996, and is currently $130,000. The Commission's order-issuing authority under Section 161 of the Atomic Energy Act is broad and extends to any area of licensed activity that affects public health and safety or the common defense and security. Orders modify, suspend, or revoke licenses, or they may require specific actions by licensees or persons. The NRC issues notices of violations and civil penalties on the basis of violations. Orders may be issued for violations or, in the absence of a violation, because of a public health or safety or common defense and security issue.

After identifying a violation, the NRC assesses its significance by considering the following factors:

- actual safety consequences
- potential safety consequences
- potential for impacting the NRC's ability to perform its regulatory function
- any willful aspects of the violation

Given those factors, the NRC takes one of the following actions based on the significance of the violation:

---

[2]   The NRC also uses administrative actions, such as notices of deviation, notices of nonconformance, confirmatory action letters, and demands for information to supplement its enforcement program.

- assigns a severity level, ranging from Severity Level IV (more than minor concern) to Severity Level I (the most significant)

- associates the violation with findings assessed through the Reactor Oversight Process significance determination process (described in Article 6) and assigns a color code of green, white, yellow, or red based on increasing risk significance

The Commission recognizes that there are violations of minor safety or environmental concern that are below Severity Level IV violations, as well as below violations associated with green findings. These minor violations are not assigned a severity level category or a color assessment.

The NRC may hold a predecisional enforcement conference or a regulatory conference with a licensee before making an enforcement decision if (1) escalated enforcement action appears warranted, (2) the NRC decides a conference is necessary, or (3) the licensee requests it. The purpose of the conference is to obtain information to assist the NRC in determining the appropriate enforcement action, such as a common understanding of facts, root causes, and missed opportunities associated with the apparent violations; corrective actions taken or planned; and the significance of issues and the need for lasting, comprehensive corrective actions.

At several junctions during the enforcement process involving cases of discrimination or willful violation of NRC regulations, the agency offers its licensees or individuals the opportunity to participate in the Alternative Dispute Resolution Program. Alternative dispute resolution is a general term encompassing various techniques for resolving conflict outside of court using a neutral third party. The NRC uses mediation, a technique in which a neutral mediator with no decisionmaking authority helps parties clarify issues, explore settlement options, and evaluate how best to advance their respective interests. Neutral mediators are selected from a roster of experienced mediators provided by a neutral program administrator who is under contract with the NRC. The mediator's responsibility is to assist the parties in reaching an agreement. However, the mediator has no authority to impose a resolution upon the parties. Mediation is a confidential and voluntary process. If the parties to the process (the NRC and the licensee or individual) agree to use alternative dispute resolution, they select a mutually agreeable neutral mediator and share equally the cost of the mediator's services. In cases in which the NRC and the licensee or the individual reach an agreement, the agency issues a confirmatory order reflecting the terms of the agreement.

The agency normally assesses civil penalties for Severity Level I and II violations, as well as knowing and conscious violations of the reporting requirements of Section 206 of the Energy Reorganization Act. Civil penalties are considered for Severity Level III violations. Although not normally used for violations associated with the Reactor Oversight Process, civil penalties (and the use of severity levels) are considered for issues that are willful, that have the potential to affect the regulatory process, or that have actual consequences.

Although each severity level may have several associated considerations, the outcome of the assessment process for each violation or problem (absent the exercise of discretion) results in one of three outcomes, which may involve no civil penalty, a base civil penalty, or twice the base civil penalty.

The NRC may issue orders to modify, suspend, or revoke a license; issue orders to cease and desist from a given practice or activity; or take such other action as may be proper. The agency may issue orders in lieu of, or in addition to, civil penalties. Additionally, the NRC may issue an order to impose a civil penalty when a licensee refuses to pay a civil penalty or an order to an unlicensed person (including vendors) when the agency has identified deliberate misconduct. By statute, a licensee or individual may request a hearing upon receiving an order. Orders are normally effective after a licensee or individual has had an opportunity to request a hearing (i.e., 30 days). However, orders can be made immediately effective without prior opportunity for a hearing when the agency determines it is the best interest of public health and safety to do so. Subsequent to the hearing process, a licensee or individual may appeal the administrative hearing decision to the Commission and, if desired, appeal the Commission's decision to a U.S. court of appeals.

Providing interested stakeholders with enforcement information is very important to the NRC. Conferences that are open to public observation are included in the listing of public meetings on NRC's public Web site. The agency issues a press release for each proposed civil penalty or order. All orders are published in the *Federal Register*. Significant enforcement actions (including actions to individuals) are included in the enforcement document collection in the Electronic Reading Room of NRC's public Web site.

From October 1, 2004, through September 30, 2005, the NRC issued a variety of significant enforcement actions to operating power reactors. Specifically, these actions included 23 escalated notices of violation without civil penalties, ten civil penalties, one confirmatory order, and one order imposing a civil monetary penalty.

During 2006, NRC issued a variety of significant enforcement actions to operating power reactors. Specifically, these actions included 26 escalated notices of violation without civil penalties, three civil penalties, and two orders.

To provide accurate and timely information to all interested stakeholders and enhance the public's understanding of the enforcement program, the NRC publishes related information on NRC's public Web site.

ARTICLE 10. PRIORITY TO SAFETY

Each Contracting Party shall take the appropriate steps to ensure that all organizations engaged in activities directly related to nuclear installations shall establish policies that give due priority to nuclear safety.

NRC policies that give due priority to safety covered under this article are PRA policy statements and policies that apply to licensee safety culture and safety culture at the NRC.

Other articles (e.g., Articles 6, 14, 18, and 19) also discuss activities undertaken to achieve nuclear safety at nuclear installations.

This section was updated to discuss new regulations, developments in PRA, and safety culture.

10.1 Background

The United States has made substantial progress in developing and using the results of PRAs for all operating reactor facilities, and the NRC has developed extensive guidance regarding the role of PRA in U.S. regulatory programs. The NRC has extensively applied information gained from PRA to complement other engineering analyses in improving issue-specific safety regulation and in changing the current licensing bases for individual plants. The move toward risk-informing the current regulations and processes continues to mark perhaps the most significant changes at the NRC. An example since the third review meeting is 10 CFR 50.69, "Risk-Informed Categorization and Treatment of Structures, Systems, and Components," promulgated on November 22, 2004. This new, voluntary rule modifies the scope of the special treatment regulations in 10 CFR Part 50. It does so by creating an alternative regulatory framework that enables licensees to use a risk-informed approach to categorize SSCs, and their associated treatment, according to their safety significance. The NRC is also continuing a program to develop additional changes to the specific technical requirements in the body of 10 CFR Part 50.

The NRC is considering an approach that, in addition to the ongoing effort to revise some specific regulations to be risk-informed and performance-based, would establish a comprehensive set of risk-informed and performance-based requirements applicable for all nuclear power reactor technologies as an alternative to current requirements. This new rule (10 CFR Part 53) would take advantage of operating experience, lessons learned from the current rulemaking activities, advances in the use of risk-informed technology, and would focus NRC and industry resources on the most risk-significant aspects of plant operations to better ensure public health and safety. The set of new alternative requirements would be intended primarily for new power reactors although they would be available to existing reactor licensees. On May 4, 2006, the NRC issued an Advanced Notice of Proposed Rulemaking seeking stakeholder input and currently is evaluating that input.

10.2 Probabilistic Risk Assessment Policy

Three policy statements form the basis for the NRC's current treatment of PRA and the related regulatory safety goals and objectives—the "Policy Statement on Severe Reactor Accidents Regarding Future Designs and Existing Plants," issued August 8, 1985; the "Safety Goals for the

Operations of Nuclear Power Plants Policy Statement," issued August 21, 1986; and the "Policy Statement on Use of PRA Methods in Nuclear Activities," issued August 16, 1995. Previous U.S National Reports have detailed these policies.

10.3 Applications of Probabilistic Risk Assessment

The NRC applies PRA to resolve severe accident issues, evaluate new and existing requirements and programs, implement risk-informed regulation, and improve data and methods of risk analysis. The NRC also engages in cooperative activities with industry (such as pilot programs for 10 CFR 50.69, 10 CFR 50.48(c), and Regulatory Guide 1.200) and in activities that assess risk in determining plant-specific changes to the licensing basis.

The NRC maintains a risk-informed and performance-based plan (RPP), which sets forth the agency's planned actions to risk-inform and performance-base its regulatory activities. In the past, the Risk-Informed Regulation Implementation Plan focused largely on risk-informed initiatives. In the current improved plan, the objectives have been expanded to more fully achieve a risk-informed and performance-based regulatory structure. A website will be developed for the RPP with links to documents that specifically describe activities and status.

The NRC and industry representatives have cooperated in a number of activities and pilot programs to develop and apply risk-informed methodologies for specific regulatory applications. The staff uses the lessons learned from these activities to enhance the effectiveness of developed guidance. The activities described in this section are inservice testing, inservice inspection, technical specification changes, and standards development.

10.3.1 Risk-Informed Inservice Testing

The agency has approved or is reviewing several applications of risk-informed inservice testing, of generally limited scope. For example, in August 2001, the staff granted a risk-informed exemption request from the licensee of the South Texas Project regarding special treatment requirements of low-risk and nonrisk-significant safety-related nuclear components (including an exemption from prescriptive inservice testing requirements). Having successfully implemented this exemption, the staff developed a new rule, 10 CFR 50.69 (discussed in Section 10.1 of this report), to allow risk insights to be applied to reduce the special treatment requirements in 10 CFR Part 50 for SSCs that are categorized as being of low safety significance.

As another example, the NRC amended 10 CFR 50.55a, "Codes and Standards," by publishing a final rule, entitled "Incorporation by Reference of ASME BPV and OM Code Cases." This rulemaking incorporated by reference specific revisions of NRC Regulatory Guides 1.84, "Design, Fabrication, and Materials Code Case Acceptability," issued March 2003; 1.147, "Inservice Inspection Code Case Acceptability, ASME Section XI, Division 1," issued January 2004; and 1.192, "Operations and Maintenance Code Case Acceptability, ASME OM Code," issued June 2003, which list the ASME Code Cases that the NRC accepts as alternatives to ASME Code requirements. Since Regulatory Guide 1.192 lists acceptable (and conditionally acceptable) OM Code Cases, including risk-informed categorization and component-specific code cases, licensees can now implement risk-informed inservice testing programs without following Regulatory Guide 1.175 and without prior NRC approval.

10.3.2 Risk-Informed Inservice Inspection

The NRC uses the guidance in Revision 1 to Regulatory Guide 1.178, "An Approach for Plant Specific Risk-Informed Decision-making for Inservice Inspection of Piping," issued September 2003, and the corresponding Section 3.9.8 of NUREG-0800. The agency-approved industry methodologies, developed by the Westinghouse Owners Group and the other by EPRI, regarding alternatives to the ASME XI Inservice Inspection Program continue to be used for inservice inspections.

The NRC regularly participates in the ASME Code development process to resolve issues regarding risk-informed inservice inspection methodology.

10.3.3 Risk-Informed Technical Specification Changes

Since the mid-1980s, the NRC has reviewed and granted improvements to technical specifications that are based, at least in part, on PRA insights. In its final policy statement on technical specification improvements published on July 22, 1993, the Commission stated that it expects licensees to use any plant-specific PRA or risk survey in preparing submittals related to technical specifications. The Commission reiterated this point when it revised 10 CFR 50.36, "Technical Specifications," in July 1995.

The NRC continues to use Regulatory Guide 1.177, "An Approach for Plant-Specific, Risk-Informed Decisionmaking: Technical Specifications," issued August 1998, and a companion section of NUREG-0800 to guide licensees on making risk-informed changes to plant technical specifications. The agency uses this regulatory guide as well as Regulatory Guide 1.174 to improve plant technical specifications. The industry and the NRC continue to increase the use of PRA in developing improvements to technical specifications. The following summarizes the major accomplishments in this area:

- Initiative 1, "Modified End States": This initiative would allow (following a risk assessment) some equipment to be repaired during hot shutdown rather than cold shutdown. The topical reports supporting this initiative for both boiling water reactor and Babcock & Wilcox plants have been approved. The Westinghouse topical report, submitted September 2005, is under review.

- Initiative 4b, "Risk-Informed Completion Times": The overall objective of this initiative is to modify technical specifications to reflect a configuration risk management approach that is more consistent with the approach of the maintenance rule (10 CFR 50.65(a)(4)). Draft industry guidance and pilot plant applications are undergoing review. The South Texas Project review will be completed in FY 2007. The Fort Calhoun Station review will be completed in FY 2008.

- Initiative 5b, "Risk informed method for Control of Surveillance Frequencies": This initiative would allow licensees to modify the frequency of technical specification surveillances based on test and risk data. The staff approved the use of this initiative for Limerick Generating Station as a pilot plant in September 2006. The staff is currently reviewing industry generic guidance that may be adopted by numerous licensees in beginning in the summer of 2008.

- Initiative 6, "Modification of Limiting Condition for Operation 3.0.3, "Actions and Completion Times": The industry is in the process of resolving discrepancies between its Combustion Engineering topical report and the NRC's Topical Report Safety Evaluation. The revised documents are expected to be resubmitted to the NRC in middle of 2008. A BWR topical report is projected to be submitted by September 2007.

- Initiative 7, "Non-Technical Specifications Support System Impact in Technical Specifications System Operability": This initiative would permit a risk-informed delay time before entering Limiting Condition for Operation actions for inoperability attributable to a loss of support function provided by equipment. Guidance documents have been approved for snubbers and hazard barriers and the industry is preparing additional proposals.

- Initiative 8, "Remove/Relocate non-safety and non-risk significant systems from Technical Specifications": This initiative would review technical specifications to remove certain system functions that had been included solely because they were judged to risk significant at one time but could be shown by additional analysis not to be. The industry and staff are in preliminary discussions on this initiative.

10.3.4 Development of Standards

The NRC worked with ASME to develop a national consensus standard for PRA related to internal initiating events. ASME issued the standard, ASME-RA-5-2002, in April 2002. ASME published the first two addenda to the standard in December 2003 and December 2005.

In parallel, the staff worked with the American Nuclear Society (ANS) to develop companion standards covering PRAs for external events, fire, low power, and shutdown operations. Work on these standards is progressing.

In January 2007, the NRC published Revision 1 to Regulatory Guide 1.200, "An Approach for Determining the Technical Adequacy of Probabilistic Risk Assessment Results for Risk-Informed Activities." This revision incorporated lessons learned from licensee-piloted risk-informed applications that rely on the regulatory guide. The agency plans further revisions to the regulatory guide to incorporate the ANS standards as they are published.

10.4 Safety Culture

An important means to implement any policy that gives due priority to safety is to foster a strong safety culture in the organization. The following discussion focuses upon safety culture, and efforts to improve safety culture, in the NRC and in the nuclear industry.

10.4.1 NRC Monitoring of Licensee Safety Culture

This section covers the policies, programs, and practices that apply to licensee safety culture.

10.4.2.1 Background

Section 6.3.2 of this report describes the Reactor Oversight Process. Based on lessons learned from the Davis-Besse reactor pressure vessel head degradation event, and other considerations, the NRC enhanced the Reactor Oversight Process to more fully address safety culture to identify safety culture problems earlier so that corrective steps can be taken to address the problems and prevent further plant performance degradation.

10.4.2.2 Enhanced Reactor Oversight Process

The NRC has adopted the IAEA International Nuclear Safety Advisory Group's definition of safety culture provided in Safety Series No.75-INSAG-4, "Safety Culture," issued February 1991, as "that assembly of characteristics and attitudes in organizations and individuals which establishes that, as an overriding priority, nuclear safety issues receive the attention warranted by their significance."

On the basis of a review of safety culture attributes developed and/or applied by the IAEA, the Nuclear Energy Agency, INPO, regulatory bodies in other countries, other domestic organizations, staff expertise, and input and feedback from NRC stakeholders, the staff identified the following components as being important to safety culture:

- decisionmaking
- resources
- work control
- work practices
- corrective action program
- operating experience
- self- and independent assessments
- environment for raising safety concerns
- preventing, detecting, and mitigating perceptions of retaliation
- accountability
- continuous learning environment
- organizational change management
- safety policies

Each one of the safety culture components is defined in a greater level of detail (e.g., cross-cutting aspects) within the Reactor Oversight Process inspection guidance documents. The Reactor Oversight Process applies the safety culture components, and their associated aspects, in different ways. The first nine safety culture components are applied in the baseline inspection and assessment program. All 13 safety culture components are applied in selected baseline, event followup, and supplemental inspection procedures.

Licensees are performing periodic voluntary self-assessments of safety culture in accordance with industry guidelines. There are no regulatory requirements for licensees to perform safety culture assessments routinely. However, depending on the extent of deterioration of licensee performance, the NRC has a range of expectations regarding regulatory actions and licensee safety culture assessments, as described below.

The Reactor Oversight Process employs a graded approach in which plants that are performing in a specified manner warrant only a routine level of inspection and oversight. However, as licensee performance deteriorates, inspection and oversight become increasingly more intrusive to ensure safe plant operation. The Reactor Oversight Process safety culture enhancements continue to allow licensees to self-diagnose and implement corrective actions for their performance problems before the NRC performs followup inspections.

For most licensees (i.e., those listed in the Licensee Response column of the Reactor Oversight Process Action Matrix), the NRC performs the baseline inspection program. In the routine or baseline inspection program, the inspector will develop an inspection finding and then identify whether an aspect of a safety culture component is a significant causal factor of the finding. The inspection findings are communicated to the licensee along with the associated safety culture aspect.

The NRC revised the inspection procedure that focuses on problem identification and resolution to allow inspectors to have the option to review licensee self-assessments of safety culture. The problem identification and resolution inspection procedure also instructs inspectors to be aware of safety culture components when selecting samples. In addition, questions related to safety-conscious work environment were enhanced in the procedure.

The agency revised the event response inspection procedures to direct inspection teams to consider contributing causes related to the safety culture components as part of their efforts to fully understand the circumstances surrounding an event and its probable cause(s).

As part of the assessment process (conducted twice per year), the NRC considers the aspects of safety culture components associated with inspection findings to determine whether common themes exist at a plant. If, over three consecutive assessment periods (i.e., 18 months), a licensee has the same safety culture issue with the same common theme, the NRC may ask the licensee to conduct a safety culture self-assessment.

As licensee performance declines (Regulatory Response column of the Reactor Oversight Process Action Matrix), the inspectors, through a specific supplemental inspection procedure, verify that the licensee's root cause, extent of condition, and extent of cause evaluations for the risk-significant finding(s) appropriately considered the safety culture components.

When the licensee performance degrades further (Degraded Cornerstone column of the Reactor Oversight Process Action Matrix), the NRC expects that the licensee's root cause evaluation for the risk-significant finding(s) determined whether any safety culture component contributed to the risk-significant performance issues. If through the conduct of a specific supplemental inspection procedure, the NRC determines that the licensee did not recognize that safety culture components caused or significantly contributed to the risk-significant performance issues, the NRC may request the licensee to complete an independent assessment of its safety culture.

Finally, for licensees with more significant performance degradation (Multiple/Degraded Cornerstone column of the Reactor Oversight Process Action Matrix), the NRC will expect the licensee to conduct a third-party independent assessment of its safety culture. The NRC will review the licensee's assessment and will conduct an independent assessment of the licensee's safety culture via a specific supplemental inspection procedure that was substantially revised to provide guidance for these assessments. The staff will be applying this revised inspection procedure for the first time at the Palo Verde plant in the latter part of 2007.

In July 2006, the NRC implemented revisions to the Reactor Oversight Process inspection and assessment processes related to safety culture. The NRC inspectors were trained on safety culture in general and on the changes to the Reactor Oversight Process before implementation. Ongoing inspector training now includes safety culture topics. The NRC is reviewing the changes to the Reactor Oversight Process over an 18-month initial implementation period that began in July 2006 and will report on the implementation lessons learned in 2008.

The safety culture changes made to the Reactor Oversight Process were intended to provide the NRC staff with (1) better opportunities to consider safety culture weaknesses and to encourage licensees to take appropriate actions before significant performance degradation occurs, (2) a process to determine the need to specifically evaluate a licensee's safety culture after performance problems have resulted in the placement of a licensee in the Degraded Cornerstone column of the Reactor Oversight Process Action Matrix, and (3) a structured process to evaluate the licensee's safety culture assessment and to independently conduct a safety culture assessment for a licensee in the Multiple/Repetitive Degraded Cornerstone column of the action matrix.

By using the existing Reactor Oversight Process framework, the NRC 's safety culture oversight activities are based on a graded approach and remain transparent, understandable, objective, risk informed, performance based, and predictable.

10.4.2 The NRC Safety Culture

The NRC has programs to promote an open, collaborative working environment that encourages all employees and contractors to promptly voice differing views without fear of retaliation. Additionally, the agency assesses the effectiveness of these programs and the working environment through the triennial Safety Culture and Climate Survey.

In fall 2005, the NRC's Inspector General (IG), with the assistance of a contractor research firm, conducted the Safety Culture and Climate Survey (see OIG-06-A-08, "OIG 2005 Survey of NRC's Safety Culture and Climate," February 10, 2006). The agency conducted similar surveys in 2002 and 1998. All NRC employees and managers were offered the opportunity to take the survey, and 70 percent responded, as compared to 53 percent in 2002. The evaluation report of the survey results is on the NRC's public Web site.

The purposes of the 2005 Safety Culture and Climate Survey were to (1) gauge the NRC's safety culture and climate, (2) compare results against the 2002 survey, and (3) compare results to Government and national benchmarks. This triennial assessment allows the NRC to address identified areas for improvement and maintain or improve areas of strength. Survey questions were grouped into 18 categories representing the major topics of the NRC's safety culture and climate. The categories singled out for particular comment are summarized below.

Compared to other benchmarks, the 2005 NRC survey results are statistically more favorable in 12 of 16 categories compared to the U.S. national norm and in 14 of 16 categories compared to the U.S. research and development norm. No categories in the survey results when compared to either of the norms are statistically less favorable. Comparing the 2005 to the 2002 NRC results, 16 of 18 categories have statistically improved, from 13 points in Communication to 3 points in Clarity of Responsibilities. The remaining two of 18 categories did not experience a statistically significant change from the 2002 results.

Compared to the 2002 survey, the NRC improved in essentially all areas, with the largest gains realized in communication, mission and strategic planning, employee engagement, recruiting, developing and retaining staff, and management leadership. Areas for improvement were workload and stress, knowledge transfer, and the use of the Differing Professional Opinion Program to raise concerns.

The NRC is acting to address the survey responses. Examples of the agency's responses to two specific areas for improvement, knowledge transfer and differing professional opinions, follow. The agency's Knowledge Management Initiative, inaugurated soon after the 2005 survey, provides a cross-office effort to capture and transfer knowledge through a variety of tools, such as formal and informal training, databases, electronic reading rooms, interviews, procedures, desk references, communities of practice, and Web sites. The NRC considers KM to be part of the strategic management of human capital, along with strategic workforce planning, recruitment, and training and development. SECY-06-0164, "NRC Knowledge Management Program," dated July 24, 2006, discusses this issue further.

The NRC's Differing Professional Opinion Program, as defined in Management Directive 10.159, "The NRC Differing Professional Opinions Program," dated May 16, 2004, allows NRC employees to bring safety concerns or other issues important to the agency's mission to senior NRC management and, where appropriate, to the Commission. This program supports the agency policy to maintain a working environment that encourages employees to make known their best professional judgments even though they may differ from the prevailing staff view, disagree with a management decision or policy position, or take issue with a proposed or established agency practice involving technical, legal, or policy issues. The program ensures the full consideration and timely disposition of differing professional opinions by affording an independent, impartial review by knowledgeable personnel who review the concern to determine the need for regulatory action. The program may be used without fear of retaliation, pressure, penalty, or unauthorized divulgence.

On November 29, 2006, the Executive Director for Operations issued draft Management Directive and Handbook 10.158, "NRC Non-Concurrence Process," as interim policy to provide agencywide instructions and guidance on raising concerns during the document concurrence process. One of this new program's objectives is to promote discussion and consideration of differing views on documents in the concurrence process before the prevailing staff view is fully developed or management or policy decisions are made.

These programs, along with the NRC's open door policy, affirm the agency's commitment to, and expectations for, an open, collaborative working environment. Employees are reminded of their responsibility to raise concerns as early as possible, engage in discussions, and seek solutions using one of the above approaches.

ARTICLE 11. FINANCIAL AND HUMAN RESOURCES

1.   **Each Contracting Party shall take the appropriate steps to ensure that adequate financial resources are available to support the safety of each nuclear installation throughout its life.**

2.   **Each Contracting Party shall take the appropriate steps to ensure that sufficient numbers of qualified staff with appropriate education, training, and retraining are available for all safety-related activities in or for each nuclear installation, throughout its life.**

This section explains the requirements regarding the financial resources that licensees must have to support the nuclear installation throughout its life and the regulatory requirements for qualifying, training, and retraining personnel.

The NRC updated this section to incorporate changes in the dollar amounts for liability under the Price-Anderson Act and thoroughly revised section 11.2.

11.1 Financial Resources

Adequate funds for safe construction, operation, and decommissioning are necessary for the protection of public health and safety. Although there does not appear to be a consistent relationship between a licensee's finances and operational safety, some evidence suggests that financial pressures have limited the resources devoted to corrective actions, plant improvements, and other safety-related expenditures. Furthermore, because a power reactor must operate to provide revenues for eventual plant decommissioning, any shutdown of a plant before its owner has accumulated sufficient funds for decommissioning could potentially hinder the safe decommissioning of that plant.

Additionally, many States (i.e., of the United States) have initiated or completed actions to economically deregulate their nuclear power plants. Traditionally, nuclear power plant owners in many States have been large, vertically integrated companies with substantial assets in generation, transmission, and distribution. In exchange for having exclusive franchises to provide electric power in defined geographical areas, nuclear plant owners have had the rates they charge to their customers regulated by State governmental bodies. This system of rate-based regulation has ensured a source of funds for construction, operation, and decommissioning of nuclear power plants. Nonetheless, this model of rate-based regulation has been changing and the United States has adjusted its processes.

The NRC distinguishes between financial qualifications for construction and operation, as well as decommissioning of nuclear power plants, and has separate regulations and programs that apply to each. The NRC also implements programs to ensure that the public has financial protection for bodily injury and property damage losses in the event of an accident. Finally, the agency has implemented requirements to ensure that licensees have insurance to help pay onsite recovery costs resulting from accidents and to provide funds for postaccident restart or decommissioning.

11.1.1 Financial Qualifications Program for Construction and Operations

This section explains the financial qualifications program for construction and operations. It explains NRC reviews for construction permits, operating licenses, combined licenses, postoperating nontransferred licenses, and license transfers.

Section 182.a of the Atomic Energy Act provides that "each application for a license…shall specifically state such information as the Commission, by rule or regulation, may determine to be necessary to decide such of the technical and financial qualifications of the applicant…as the Commission may deem appropriate for the license." To implement this provision, the NRC has developed the regulations and guidance discussed below.

11.1.1.1 Construction Permit Reviews

As required by 10 CFR 50.33(f)(1), applicants for construction permits must submit information that "demonstrates that the applicant possesses or has reasonable assurance of obtaining the funds necessary to cover estimated construction costs and related fuel cycle costs." Appendix C, "A Guide for the Financial Data and Related Information Required to Establish Financial Qualifications for Facility Construction Permits," to 10 CFR Part 50 provides more specific directions for evaluating the financial qualifications of applicants.

11.1.1.2 Operating License Reviews

An "electric utility" as defined in 10 CFR 50.2, "Definitions," is "any entity that generates or distributes electricity and which recovers the cost of this electricity, either directly or indirectly, through rates established by the entity itself or by a separate regulatory authority." Electric utilities are exempt under 10 CFR 50.33(f) from reviews of financial qualifications of applications for operating licenses. The reason for this exemption is that cost-of-service rate regulation, as it has existed in the United States, has ensured that ratepayers provide a source of funds for the safe operation of nuclear power plants. Applicants for operating licenses that are not electric utilities are required under 10 CFR 50.33(f)(2) to submit information that demonstrates that they possess or have reasonable assurance of obtaining the necessary funds to cover estimated operating costs. Nonelectric-utility applicants for operating licenses are also required to submit estimates for the total annual operating costs for each of the first 5 years of operation of their facilities and must indicate the sources of funds to cover operating costs.

11.1.1.3 Combined License Application Reviews

As authorized in 10 CFR Part 52, applicants may apply for a combined construction permit and operating license. Under 10 CFR 52.77, "Contents of Applications; Technical Information," such applications must contain all of the information required under 10 CFR 50.33, including information regarding financial qualifications. The procedures described above are used to review future combined license applications that the NRC receives.

11.1.1.4 Postoperating License Nontransfer Reviews

The NRC does not systematically review the financial qualifications of power reactor licensees once it has issued an operating license, other than for license transfers as described below.

However, as provided in 10 CFR 50.33(f)(4), the NRC can seek additional information on licensees' financial resources if the agency considers such information appropriate.

11.1.1.5 Reviews of License Transfers

The NRC regulations in 10 CFR 50.80 require agency review and approval of transfers of operating licenses, including licenses for nuclear power plants that are owned or operated by electric utilities. The NRC performs these reviews to determine whether a proposed transferee or new owner is technically and financially qualified to hold the license.

NUREG-1577, Revision 1, "Standard Review Plan on Power Reactor Licensee Financial Qualifications and Decommissioning Funding Assurance" describes the agency's overall review process for applicant and licensees' financial qualifications for nuclear power plant construction and operation.

11.1.2 Financial Qualifications Program for Decommissioning

Among other sections of the Atomic Energy Act, Section 182.a establishes the basis for NRC's decommissioning funding assurance regulations and guidance; the previous U.S. National Report listed specific regulations and guidance.

11.1.3 Financial Protection Program for Liability Claims Arising from Accidents

The Price-Anderson Act, enacted in 1957, became Section 170 of the Atomic Energy Act and governs the U.S. financial protection program. Section 170 (with related definitions in Section 11) provides the financial and the legal framework to compensate those who suffer bodily injury or property damage as a result of accidents at nuclear facilities covered by the law's provisions. The NRC regulations implementing the provisions of Section 170 for NRC licensees are codified in 10 CFR Part 140, "Financial Protection Requirements and Indemnity Agreements."

The Price-Anderson Act was enacted to (1) remove the deterrent to private sector participation in atomic energy presented by the threat of potentially enormous liability claims in the event of a catastrophic nuclear accident and (2) ensure that adequate funds are available to the public to satisfy liability claims if such an accident were to occur.

The Price-Anderson Act was revised most recently in 2005, when Congress renewed the Commission's authority to cover new facilities until 2020. Under current law, power reactors over 100 megawatts electric must contribute to a pool that replaces the Government as the second provider of funds if the first layer of financial protection (liability insurance—now $300 million) is exhausted.

The reactor operators are required after an accident to pay into a "retrospective premium pool," in maximum annual installments not to exceed $15 million, up to a total of $95.8 million each. But payment is called for only if the accident exhausts the first layer of financial protection, and only if, and to the extent that, additional funds are needed to pay the damages. With 104 reactors currently participating in the system, the total financial protection available under the Price-Anderson Act for any one accident is $10 billion ($300 million primary coverage + (95.8 million per reactor × 104 reactors)); $10 billion is also the limit on liability. As reactors leave the retrospective premium system as a result of permanent closure or join as the result of construction of new reactors, this coverage limit may fall or rise. A change in the limit may also occur when the $95.8 million contribution is adjusted for inflation, as must be done every 5 years. In any event, Congress will address any damages exceeding the total sum that reactors must contribute to the pool and will decide upon the next steps needed for compensation.

The public is significantly benefitted by another feature of the act. Claimants need only prove that the accident caused their injury in order to receive compensation for damages from any accident with significant offsite releases of radiation (i.e., an "extraordinary nuclear occurrence"). Neither proof of fault nor proof of what caused the accident is necessary.

Claims for more than 150 alleged incidents involving nuclear material have been filed under various liability policies since the inception of the Price-Anderson Act in 1957. The insured losses and expenses paid so far total more than $125 million. Of this amount, most payments arose out of the accident at TMI Unit 2.

11.1.4 Insurance Program for Onsite Property Damages Arising from Accidents

Among other sections of the Atomic Energy Act, Section 182.a provides the basis for the NRC's onsite property damage insurance requirements for operating nuclear power reactors contained in 10 CFR 50.54(w).

The U.S. nuclear industry has not experienced an accident involving significant radioactive contamination since TMI Unit 2.

11.2 Regulatory Requirements for Qualifying, Training, and Retraining Personnel

This section explains the regulatory requirements for qualifying, training, and retraining personnel. It discusses the governing documents, the process for implementing requirements, and experience. It also discusses the INPO accreditation activities.

11.2.1 Governing Documents and Process

The training requirements for licensed operators and licensed senior operators are regulated by 10 CFR Part 55, "Operators' Licenses," which allows facility licensees to have operator requalification program content that is derived using a systems approach to training (SAT; as defined in 10 CFR 55.4) or that meets the requirements outlined in 10 CFR 55.59(c). Subpart D, "Applications," of 10 CFR Part 55 requires that operator license applications must contain information about an individual's training and experience, unless the facility licensee certifies that the applicant has successfully completed a Commission-approved training program that is SAT-based and uses an acceptable simulation facility.

The operator licensing process at power reactors includes a generic fundamentals examination covering the theoretical knowledge that is required to operate a nuclear power plant; license applicants must pass the generic fundamentals examination before they can take a site-specific examination. The site-specific examination consists of a written examination and an operating test that includes a plant walkthrough and a dynamic performance demonstration on a simulation facility.

The NRC staff has transferred most of the responsibility for developing site-specific licensing examinations to facility licensees. In 1999, the NRC amended 10 CFR Part 55 to allow nuclear power reactor licensees to prepare the written examinations and operating tests that the agency uses to evaluate the competence of applicants for operators' licenses at those facilities. Licensees that elect to prepare their own examinations are required to establish procedures to control examination security and integrity. The licensees prepare and submit proposed examinations and operating tests to the NRC according to the guidance in NUREG-1021, "Operator Licensing Examination Standards for Power Reactors," Revision 9, issued July 2004. The NRC reviews the facility-prepared examinations, prepares examinations for facility licensees upon request, administers all operating tests, makes the final licensing decisions, and issues the licenses.

As required by 10 CFR 50.120, "Training and Qualification of Nuclear Power Plant Personnel," training programs must be established, implemented, and maintained using a SAT approach for eight categories of nonlicensed workers at nuclear power plants and the shift supervisor, who is licensed in accordance with 10 CFR Part 55. These provisions complement the requirements for training based on a systems approach for the requalification of licensed operators and licensed senior operators. Regulatory Guide 1.8, "Qualification and Training of Personnel for Nuclear Power Plants," Revision 3, issued May 2000, contains guidance to implement the regulations.

The NRC continues to endorse the training accreditation process managed by INPO. The staff recognizes that training programs developed in accordance with INPO guidelines and accredited by the National Nuclear Accrediting Board (NNAB) are SAT-based; therefore, accredited programs are considered to be consistent with the regulations in 10 CFR Part 55 and 10 CFR Part 50.120. The NRC also recognizes that INPO-managed accreditation and associated training evaluation activities are an acceptable means of self-improvement in training. Such recognition encourages industry initiative and reduces NRC evaluation and inspection activities.

In accordance with its memorandum of agreement with INPO, the NRC monitors INPO accreditation activities as part of its continuing assessment of the effectiveness of the industry's training programs. Specifically, the NRC staff observes selected accreditation team visits and NRC managers periodically observe NNAB meetings. These observations are intended to monitor the implementation of programmatic aspects of the accreditation process, but they also provide an opportunity to assess the selected performance areas of facility licensees.

If the NNAB has concerns regarding the performance of an accredited training program, it will place the program on probation. This does not necessarily place a training program in noncompliance with either 10 CFR Part 55 or 10 CFR 50.120, since training programs are accredited to a standard of excellence rather than a minimum level of regulatory compliance. However, the NRC does review the circumstances leading to the probation to ensure safe operations and continued compliance with the regulations.

The NNAB may also withdraw accreditation in response to major deficiencies in a licensee's accredited training program. If accreditation is withdrawn, the NRC would request that the licensee report the circumstances of the withdrawal for the staff to determine the significance of the issues related to the withdrawal. If the NRC determines that compliance with the regulations is not affected, it may not be necessary to take any further action. If the withdrawal is linked to a breakdown in the training process or a safety-significant issue, the NRC will conduct an immediate inspection focused on the process problem or safety issue(s). The agency would take further action, such as issuing confirmatory action letters or orders, if appropriate.

The NRC monitors industry performance in implementing the training requirements of 10 CFR Part 50 and 10 CFR Part 55 by (1) reviewing licensee event reports and inspection reports for training issues, (2) observing the accreditation process, and (3) reviewing the results of operator licensing activities. Guidance for periodically inspecting the licensed operator requalification training program at every facility is given in Inspection Procedure (IP) 71111.11, "Licensed Operator Requalification Program." In addition, when appropriate for cause, the NRC will use IP 41500, "Training and Qualification Effectiveness," which references the guidance in NUREG-1220, "Training Review Criteria and Procedures," Revision 1, issued January 1993, to verify compliance with SAT requirements.

11.2.2 Experience

The NRC reviewed training issues contained in licensee event reports and inspection reports during 2006 using data from the Human Factors Information System, which is described in Article 12. The review revealed that the proportion of human performance issues attributable to training, for U.S. nuclear power plants, decreased from eight percent in 1999 to four percent in 2005. The training-related issues identified were concentrated in two subcategories, "training less than adequate" and "individual knowledge less than adequate." The NRC annually assesses the effectiveness of training in the nuclear industry and prepares a report of its findings; the reports for 1999 through 2005 are available on the NRC's public Web site.

Although the NRC identified some limited specific weaknesses in training programs, all indicators suggest that the industry is successfully implementing training programs in accordance with the regulations. The NRC will continue to monitor selected performance areas, emphasizing the identification and resolution of training process problems.

An example of this monitoring process is the for-cause training inspection conducted at Virgil C. Summer Nuclear Station in May 2006. The NRC performed this inspection in response to the high failure rate on the written portion of the initial operator license examination administered in January 2006. The inspection provided reasonable assurance that the operator license applicants had successfully completed the initial training program and that the program meets SAT requirements.

ARTICLE 12. HUMAN FACTORS

Each Contracting Party shall take the appropriate step to ensure that the capabilities and limitations of human performance are taken into account throughout the life of a nuclear installation.

This section explains the NRC program on human performance. The seven major areas under this program are (1) human factors engineering issues, (2) emergency operating procedures and plant procedures, (3) working hours and staffing, (4) fitness-for-duty, (5) human factors information system, (6) support to event investigations and for-cause inspections, and (7) training.

The NRC updated this section to incorporate new documents, issues, and experience.

12.1 Goals and Mission of the Program

The NRC has a comprehensive program for ensuring that human performance is properly addressed in a risk-informed regulatory framework for maintaining reactor safety. The NRC developed the program based on reviewing risk information and activities in the domestic and international nuclear industry.

12.2 Program Elements

The Reactor Oversight Process (discussed in Article 6) focuses on cornerstones of safety that are assessed through a combination of performance indicators and risk-informed inspections. The inspections focus on risk-significant activities and systems related to the cornerstones. The three elements that are considered cross-cutting with respect to the cornerstones are human performance, safety-conscious work environment, and corrective actions. The Human Performance Program has contributed directly to the development of a supplemental inspection procedure related to the human performance cross-cutting element. The Human Performance Program is also engaged in the other two elements, since a safety-conscious work environment and many of the actions involved in corrective action programs result from human performance problems.

The Human Performance Program also supports the Risk-Informed and Performance-Based Plan by generating, collecting, and evaluating data on human performance for use in human reliability analysis models. The staff evaluates information to gain insights to support risk-informed regulation and to provide human performance data for human reliability analysis.

The Human Performance Program monitors technological developments and emerging issues in an effort to prepare the NRC for the future. Two ongoing activities include developing regulatory guidance for reviewing designs of control stations and processing requests related to deregulation. Licensees are replacing aging analog controls and displays with digital components, and the NRC must be prepared to review safety issues with respect to human-system interfaces resulting from such new designs and technologies. With regard to deregulation, the NRC has been processing numerous industry requests to transfer operating licenses, which may involve changes in organizational structure affecting human performance.

12.3 Significant Regulatory Activities

The NRC performs significant regulatory activities in the following seven areas to address human performance:

(1)   human factors engineering Issues
(2)   emergency operating procedures and plant procedures
(3)   shift staffing
(4)   fitness-for-duty
(5)   human factors information system
(6)   support to event investigations and for-cause inspections
(7)   training

The first six activities are described below; Article 11 describes training.

12.3.1 Human Factors Engineering Issues

This section discusses human factors activities related to engineering issues, covering the governing documents and process to carry out requirements and experience.

Governing Documents and Process. The NRC evaluates the human factors engineering design of the main control room and control centers outside of the main control room using Chapter 18 of NUREG-0800, NUREG-0700, "Human System Interface Design Review Guideline," Revision 2, issued May 2002, and NUREG-0711, "Human Factors Engineering Program Review Model, Revision 2" issued February 2004. These documents provide guidance for the review of human-system interface issues in connection with the design certification of nuclear installations and the NRC's inspection program. The NRC also uses NUREG-1764, "Guidance for the Review of Changes to Human Actions," to review license amendment requests that credit the use of manual actions. Moreover, IN 97-78, "Crediting of Operator Actions in Place of Automatic Actions and Modifications of Operator Actions, Including Response Times," dated October 23, 1997, identifies references that the NRC uses to review the completion times of operator manual actions and how the actions will be reflected in the licensee's emergency procedures and operator training.

Experience. The NRC reviews licensees' requests that involve aspects of human factors engineering. Examples include crediting operator manual actions in amendments to plant technical specifications, transferring facility operating licenses, and increasing the reactor's authorized power level (i.e., power uprates). Recent license amendment requests from Fort Calhoun Station and R.E. Ginna Nuclear Power Plant are examples of NRC reviews involving new or modified operator manual actions. The amendments from Fort Calhoun proposed crediting manual actions to isolate steam generator blowdown within a certain time during a loss of main feedwater event and to reduce the number of containment spray pumps needed in response to a LOCA. The amendment request from Ginna also involved operator manual actions to realign safety injection pumps for recirculation following a LOCA.

The NRC has also evaluated a number of requests to transfer facility operating licenses, which affected management and organization, staffing, and technical qualifications. The NRC used Chapter 13 of NUREG-0800 as the principal guidance for these reviews.

The NRC also reviews and approves requests for power uprates from currently licensed plants. For such requests, the NRC examines the effect of the power uprate on plant procedures, controls, displays, and alarms, and required operator actions using Review Standard (RS-001), "Review Standard for Extended Power Uprates," Section 2.11.1 (in both BWR and PWR sections). The NRC issued this guidance in December 2003; it can be found on the NRC's public Web site along with additional general information on power uprates. The agency recently reviewed and approved power plant uprates for R.E. Ginna, Beaver Valley Units 1 and 2, and Vermont Yankee Nuclear Power Station.

12.3.2 Emergency Operating Procedures and Plant Procedures

Licensees must have programs to develop, implement, and maintain emergency operating and plant procedures. (Article 16 discusses emergency preparedness; this discussion is limited to the human factors aspect of emergency operating procedures.)

Governing Documents and Process. On December 17, 1982, the NRC issued GL 82-33, "Requirements for Emergency Response Capability," which transmitted Supplement 1 to NUREG-0737, "Requirements for Emergency Response Capability," requires each licensee to submit a set of documents for developing emergency operating procedures.

Experience. No significant examples related to emergency operating and plant procedures were identified since 2004.

12.3.3 Shift Staffing

Governing Documents and Process. Paragraph (m) of 10 CFR 50.54, "Conditions of Licenses," specifies the minimum number of licensed operators and senior operators that are required for nuclear power reactor facilities. Appendix R, "Fire Protection Programs for Nuclear Power Facilities Operating Prior to January 1, 1979," and Appendix E, "Emergency Planning and Preparedness for Protection and Utilization Facilities," to 10 CFR Part 50 contain the NRC staffing requirements for fire brigades and emergency response personnel.

In September 2002, the NRC began work on a process to evaluate exemption requests from the requirements in 10 CFR 50.54(m) resulting from the changing demands and new technologies presented by advanced reactor control room designs and significant light-water reactor control room upgrades. In July 2005, the NRC published NUREG-1791, "Guidance for Assessing Exemption Requests from the Nuclear Power Plant Licensed Operator Staffing Requirements Specified in 10 CFR 50.54(m)." The purpose for reviewing the exemption requests is to ensure public health and safety by verifying that the applicant's staffing plan and supporting analyses sufficiently justify the requested exemption. NUREG/CR-6838, "Technical Basis for Assessing Exemptions from Nuclear Power Plant Licensed Operator Staffing Requirements 10 CFR 50.54(m)," issued February 2004, explains the justification for the recommended process.

Experience. No significant examples related to shift staffing were identified for 2004–2006.

12.3.4 Fitness-for-Duty

This section discusses the NRC's requirements pertaining to the fitness-for-duty of nuclear power plant workers, including requirements regarding the control of work hours and management of worker fatigue.

Governing Documents and Process. As required by 10 CFR Part 26, "Fitness for Duty Programs," each licensee authorized to operate or construct a nuclear power reactor must implement a fitness-for-duty program for all personnel having unescorted access to the protected area of its plant. As performance objectives, 10 CFR Part 26 requires that licensees establish programs that (1) provide reasonable assurance that nuclear power plant personnel perform their tasks in a reliable and trustworthy manner and are not under the influence of any substance, legal or illegal, or mentally or physically impaired from any cause, (2) provide reasonable measures for the early detection of persons who are not fit to perform activities, and (3) have a goal of achieving a drug-free workplace and a workplace free of the effects of such substances.

The NRC issues annual reports on statistical data and lessons learned by licensees from their fitness-for-duty program performance reports. The most recent of these is NRC Information Notice 2006-30, "Summary of Fitness-for-Duty Program Performance Reports for Calendar Years 2004 and 2005," dated December 21, 2006. The NRC is currently completing a similar report that will provide data for 2006. A project to automate the reporting and trending of performance data using a Web-based approach is ongoing. In addition, the NRC has established an email address for licensees and individuals to submit fitness-for-duty questions, as well as a Web site where performance reports and the answers to frequently asked questions are publicly available.

Regarding worker fatigue, in 1982, the NRC issued its "Policy on Factors Causing Fatigue of Operating Personnel at Nuclear Reactors." The objective of the policy is for licensees to ensure, to the extent practicable, that personnel who perform safety-related functions are not assigned to shift duties while in a fatigued condition that can significantly reduce their mental alertness or decisionmaking ability. Licensees have implemented the policy through reactor facility technical specifications that require administrative control of personnel work hours. In April 2003, the NRC issued Order-03-038, "Order for Fitness-for-Duty Enhancements for Nuclear Security Force Personnel," requiring licensees to implement compensatory measures to control the work hours of security personnel at nuclear power plants. The order requires licensees to maintain the work hours of security force personnel below specified levels, monitor for individual fatigue, and establish a process to follow if an individual declares that he or she is unfit for duty as a result of fatigue. The NRC issued the order in response to concerns associated with a significant increase in the work hours of security personnel at U.S. nuclear facilities following the terrorist attacks on September 11, 2001.

Experience. When 10 CFR Part 26 was published in 1989, the Commission directed the NRC staff to continue to analyze licensee programs, assess the effectiveness of the rule, and recommend appropriate improvements or changes. In 2006, following years of stakeholder interactions, the NRC staff proposed a comprehensive amendment of the rule to update and enhance the fitness-for-duty requirements. The proposed amendment, which was provided to the Commission via SECY-06-0244, "Final Rulemaking—10 CFR Part 26—Fitness-for-Duty Programs," on December 22, 2006, includes provisions for specimen validity testing, stronger sanctions for refusing to submit to testing and attempts to circumvent the testing process, and increased legal protections for those tested. The proposed amendment would also establish requirements for the management of worker fatigue, including requirements for licensees to (1) train workers in methods for managing fatigue, (2) establish procedures for appropriately managing instances of individuals declaring they are unfit for duty because of fatigue, (3) limit the work hours of individuals who perform functions important to the protection of public health and safety, and (4) submit an annual report that describes how often the licensee authorizes individuals to exceed the work hour limits. Full implementation of the rule is expected to take at least one year from the date of publication in the *Federal Register*.

12.3.5 Human Factors Information System

Governing Documents and Process. The Human Factors Information System is designed to store, retrieve, sort, and analyze human performance information extracted from NRC inspection and licensee event reports. Initiated in 1990, this automated information management system can generate a variety of specialized reports that are not readily available from other NRC sources. In 2006, improvements were made to this system to better align the coding scheme with the Reactor Oversight Process (described in Article 6) and to enhance the system's search capabilities. The Human Factors Information System now captures information related to training, procedures and reference documents, fitness-for-duty, oversight, problem identification and resolution, communications, human-system interface and environment, and work planning and practices.

Experience. The NRC responds to stakeholder and public inquiries and data requests on this system on a regular basis. For example, inspectors use the data generated by this system in preparing inspection activities related to human performance. In addition, the NRC's Office of Research uses the data to support activities in human performance and human reliability analysis. Other NRC program offices use the data to gain insights about human performance and to monitor the frequency of human performance issues and to inform several types of reports, such as internal operating experience reports and the NRC's annual report on the effectiveness of training in the nuclear industry (discussed in Section 11.2.2 of this report). The NRC also uses a Web page to disseminate information on human performance issues at individual nuclear power plant sites.

12.3.6 Support to Event Investigations and For-Cause Inspections and Training

Governing Documents and Process. NRC staff members with human factors expertise are often included in special inspections, incident investigation team inspections, augmented team inspections, event investigations, and supplemental inspections. Human factors experts have assessed management effectiveness, procedures, training issues, staffing issues, human-machine interfaces, personnel performance issues, safety-conscious work environment, and safety culture.

For training issues, inspectors use IP 41500. For procedure issues, inspectors use IP 42001, "Emergency Operating Procedures," and IP 42700, "Plant Procedures." For baseline inspections under the Reactor Oversight Process (described in Article 6), inspectors use IP 71152, "Identification and Resolution of Problems," which is intended to establish confidence that each licensee is detecting and correcting problems in a manner that limits the risk to the public and includes a review of the licensee's safety-conscious work environment. A key premise of the Reactor Oversight Process is that weaknesses in problem identification and resolution programs will manifest themselves as performance issues that can be identified during the baseline inspection program or by crossing predetermined indicator thresholds.

For supplemental inspections, IP 95003, "Supplemental Inspection for Repetitive Degraded Cornerstones, Multiple Degraded Cornerstones, Multiple Yellow Inputs, or One Red Input," as revised in October 2006, includes requirements for the NRC staff to review the licensee's third-party safety culture assessment and independently assess the licensee's safety culture. Staff with technical expertise in human factors and safety culture will perform the safety culture activities. The first implementation of revised IP 95003 is scheduled for the Fall 2007.

Experience. In 2005, the NRC staff with human factors expertise participated in an IP 95003 inspection at the Perry Nuclear Power Plant to assess human performance at the site. The inspectors determined that a number of findings related to procedure adherence had strong human performance contributions. The NRC returned to Perry in 2006 to assess the effectiveness of the licensee's corrective actions in addressing the human performance issues identified during the previous inspection. The evaluation included observing the use of human performance error prevention tools during work activities in the plant, observing supervisors' use and reinforcement of these tools, and reviewing the licensee's performance indicators in the human performance area. Based on these activities, the inspectors concluded that the licensee's corrective actions were effective in improving human performance. The inspectors found that the licensee had adequately implemented commitments to address issues in the human performance area and evaluated the effectiveness of the corrective actions.

ARTICLE 13. QUALITY ASSURANCE

Each Contracting Party shall take the appropriate steps to ensure that quality assurance programmes are established and implemented with a view to providing confidence that specified requirements for all activities important to nuclear safety are satisfied throughout the life of a nuclear installation.

This section explains QA policy and requirements and guidance for design and construction, operational activities, and staff licensing reviews. It also describes QA programs and regulatory guidance. This section was updated to discuss a new construction program for expected new reactors.

13.1 Background

Nuclear power facilities must be designed, constructed, and operated in a manner that ensures (1) the prevention of accidents that could cause undue risk to the health and safety of the public and (2) the mitigation of adverse consequences of such accidents if they should occur. A primary means for achieving these objectives is by establishing and effectively implementing a nuclear QA program. Although a licensee may delegate aspects of the establishment or execution of the QA program to others, the licensee remains ultimately responsible for the program's overall effectiveness. Licensees perform a variety of self-assessments to validate the effectiveness of their QA program implementation. The NRC reviews descriptions of QA programs and performs onsite inspections to verify aspects of the program implementation.

13.2 Regulatory Policy and Requirements

Each applicant for a construction permit for a nuclear power plant is required by 10 CFR 50.34(a)(7) to describe its QA program in its preliminary safety analysis report. Also, each applicant for a combined license is required by 10 CFR 52.79(b) to describe its QA program in its preliminary safety analysis report. This program applies to the design, fabrication, construction, and testing of safety-related plant equipment. Each applicant for a license to operate a nuclear power plant is required by 10 CFR 50.34(b)(6)(ii) to provide a final safety analysis report that details its managerial and administrative controls to ensure safe operation. In both the preliminary and final safety analysis reports, the applicant must describe how it will satisfy the applicable requirements of Appendix B, "Quality Assurance Criteria for Nuclear Power Plants and Fuel Reprocessing Plants," to 10 CFR Part 50.

If a licensee wants to make changes in its QA program, it is required by 10 CFR 50.54(a)(3) to inform the NRC of the changes. A licensee can make changes without prior NRC approval if the changes do not reduce the commitments in the program description as accepted by the NRC.

Nuclear QA criteria apply to all activities that affect the safety-related functions of SSCs that prevent or mitigate the consequences of postulated accidents that could cause undue risk to the health and safety of the public. High-level criteria for determining which plant SSCs are safety related are provided in 10 CFR 50.2, "Definitions." Based upon these criteria, licensees' engineering organizations develop plant-specific listings of safety-related SSCs.

13.2.1 Appendix A to 10 CFR Part 50

Appendix A, "General Design Criteria for Nuclear Power Plants," to 10 CFR Part 50 details the general requirements for establishing QA controls. General Design Criterion 1, "Quality Standards and Records," contains certain requirements applying to the QA of items important to safety. The scope of items that are "important to safety" includes a subset of plant equipment that is classified as safety related. Appendix B, "Quality Assurance Criteria for Nuclear Power Plants and Fuel Reprocessing Plants," to 10 CFR Part 50 (discussed below) contains QA program requirements for safety-related SSCs. Other regulatory guidance discusses QA program controls that are appropriate for some types of nonsafety-related equipment.

13.2.2 Appendix B to 10 CFR Part 50

Appendix B to 10 CFR Part 50 outlines the QA requirements that apply to activities that affect the safety-related functions of SSCs that prevent or mitigate the consequences of postulated accidents. The appendix defines QA as all planned and systematic actions that are necessary to provide adequate confidence that SSCs will perform satisfactorily in service. Toward that end, Appendix B specifies 18 criteria that the commitments in a licensee's QA program must satisfy. These criteria cover such aspects as organizational independence, design control, procurement, document control, test control, corrective action, and audits. Appendix B also stipulates that licensees establish measures to ensure that applicable regulatory requirements, design bases, and other requirements that are necessary to ensure adequate quality are suitably included or referenced in the documents for procurement of safety-related materials, equipment, and services whether purchased by the licensee or its contractors or subcontractors. Consistent with the importance and complexity of the products or services to be provided, licensees (or their designees) are responsible for periodically verifying that suppliers' QA programs comply, as appropriate, with the applicable criteria in Appendix B and that they are effectively implemented. Additionally, as outlined in 10 CFR 21.41, the NRC staff performs inspections at vendors who supply basic components to the nuclear industry.

The requirements of Appendix B are written at a conceptual level and it was necessary for the NRC and the industry to develop consensus standards that include acceptable ways to conform to these requirements. The NRC then issued companion regulatory guides, which endorsed (with conditions, if warranted) QA codes and standards.

13.2.3 Approaches for Adopting More Widely Accepted International Quality Standards

As stated in the previous U.S. National Report, the NRC reviewed options for adopting more widely accepted international quality standards like International Organization for Standardization (ISO) Standard 9001, considering how international standards compare with the existing Appendix B framework. On the basis of this review, the NRC concluded that supplemental quality requirements would be needed when implementing ISO 9001 within the existing regulatory framework. As part of the ongoing multinational design evaluation program, the NRC will be reevaluating various international QA standards to achieve a greater degree of international convergence.

13.3 QA Regulatory Guidance

The NRC has QA guidance for design and construction, operational activities, and licensing.

13.3.1 Guidance for Design and Construction Activities

As discussed in the previous U.S. National Report, licensees apply consensus standards developed by the American National Standards Institute (ANSI) in its N45.2 series or ASME in its NQA-1 series as acceptable ways of complying with the requirements of Appendix B to 10 CFR Part 50. The NRC endorses ANSI and ASME standards through its regulatory guides. The NRC staff, as part of consensus codes and standards activities, continues to participate with ASME NQA-1 committees to revise the latest edition of the NQA-1 standard to be endorsable within the 10 CFR 50.55(a) regulatory framework. Currently, a number of licensees have implemented QA programs applying ASME NQA-1-1994.

13.3.2 Guidance for Operational Activities

As described in the previous U.S. National Report, the NRC has conditionally endorsed the consensus standard ANSI N18.7-1976, "Administrative Controls and Quality Assurance for the Operational Phase of Nuclear Power Plants" through Regulatory Guide 1.33, "Quality Assurance Program Requirements (Operations)," Revision 3, issued November 1980, as complying with the requirements of Appendix B.

13.3.3 Guidance for Staff Reviews for Licensing

NUREG-0800 guides the staff review of applications. NUREG-0800 has specific review guidance correlated with the 18 criteria of Appendix B to 10 CFR Part 50 and integrates a review of licensee commitments to adopt the NRC's QA-related regulatory guides and the industry's QA codes and standards. The draft of NUREG-0800, Section 17.5, provides the NRC's expectations for a QA program.

13.4 <u>QA Programs</u>

The NRC inspects QA programs under the Reactor Oversight Process for operating reactors and under the Construction Inspection Program for new reactors. Additionally, the NRC conducts augmented inspection programs as needed.

The baseline inspection program of the Reactor Oversight Process includes one primary procedure related to QA issues, known as Inspection Procedure 71152, "Identification and Resolution of Problems." Inspectors use this procedure to assess the effectiveness of licensees' programs to identify and resolve problems according to a performance-based review of specific issues. In particular, inspectors look for cases in which a licensee may have missed generic implications of specific problems and for the risk significance of combinations of problems that individually may not have significance. They also verify that licensees are properly capturing issues that could affect the availability of equipment that is tracked under 10 CFR 50.65, "Maintenance Rule," or by performance indicators. They do not inspect other aspects of QA program implementation in the baseline inspection program, but may, through supplemental inspections.

Some equipment in the nuclear facility may be classified as nonsafety related and yet still be important to safety for some unique reason. In specific cases, the NRC has specified that QA controls are warranted for equipment determined to be more important than commercial-grade equipment. However, the QA controls do not have to meet Appendix B requirements, which apply only to activities affecting safety-related functions. Typically, applying QA controls to this important-to-safety, yet nonsafety-related, equipment is called "augmented quality control."

The NRC's inspection program has been anticipating applications for combined licenses to facilitate the construction of new nuclear power plants under 10 CFR Part 52. (See Article 18.) The Construction Inspection Program provides oversight for future nuclear plants licensed under 10 CFR Part 52, including QA program inspection. The QA inspection program focuses on an applicant establishing and implementing a QA program in accordance with Appendix B to 10 CFR Part 50 to ensure compliance with specified requirements in 10 CFR Part 52.

As provided in the Construction Inspection Program (see Article 18), the plant will transition from the Construction Inspection Program to the Reactor Oversight Process for commercial operation after the completion of startup open items.

ARTICLE 14. ASSESSMENT AND VERIFICATION OF SAFETY

Each Contracting Party shall take the appropriate steps to ensure that:

(i)   comprehensive and systematic safety assessments are carried out before the construction and commissioning of a nuclear installation and throughout its life. Such assessments shall be well documented, subsequently updated in the light of operating experience and significant new safety information, and reviewed under the authority of the regulatory body

(ii)   verification by analysis, surveillance, testing, and inspection is carried out to ensure that the physical state and the operation of nuclear installations continue to be in assurance with its design, applicable national safety requirements, and operational limits and conditions

This section explains the governing documents and process for ensuring that systematic safety assessments are carried out during the life of the nuclear installation, including for the period of extended operation. It focuses on assessments performed to maintain the licensing basis of a nuclear installation. Finally, this section explains verification of the physical state and operation of the nuclear installation by analysis, surveillance, testing, and inspection.

Other articles (e.g., Articles 6, 10, 13, 18, and 19) also discuss activities undertaken to achieve safety at nuclear installations.

Updates to this section include information on the amendment process, Browns Ferry restart, and license conditions.

14.1 Ensuring Safety Assessments throughout Plant Life

Before a nuclear facility is constructed, commissioned, and licensed, an applicant must perform comprehensive and systematic safety assessments, which are reviewed and approved by the NRC. Article 18 discusses these assessments and reviews. This section focuses on the assessments that are required throughout the life of a nuclear installation (i.e., assessments required to maintain the licensing basis). To show conformance with the licensing basis, a licensee must maintain records of the original design bases and any changes. This section explains how such changes are documented, updated, and reviewed. Renewal of a license is predicated on the requirement that a licensee will continue to meet its current licensing basis; this section explains how the license renewal process accounts for this requirement.

14.1.1 Maintaining the Licensing Basis

The NRC carries out regulatory programs to give reasonable assurance that plants continue to conform to the licensing basis. (Article 6 discusses these programs.) This section explains the governing documents and process used to maintain the licensing basis. The main governing documents are 10 CFR 50.90, "Application for Amendment of License or Construction Permit," 10 CFR 50.59, "Changes, Tests, and Experiments," and 10 CFR 50.71, "Requirements for Updating of Final Safety Analysis Reports."

14.1.1.1 Governing Documents and Process

A licensee is to operate its facility in accordance with the license and as described in its final safety analysis report. To change its license or reactor facility, a licensee must follow the review and approval processes established in the regulations. For license amendments, including changes to technical specifications, the licensee must request NRC approval in accordance with 10 CFR 50.90. However, 10 CFR 50.59 (as described below) contains requirements for the process by which licensees may make changes to their facilities and procedures as described in the safety analysis report, without prior NRC approval, under certain conditions.

10 CFR 50.59. 10 CFR 50.59 establishes the conditions under which licensees may make changes to the facility or procedures and conduct tests or experiments without prior NRC approval. Proposed changes, tests, and experiments that satisfy the definitions and one or more of the criteria in the rule must be reviewed and approved by the NRC before implementation. Thus the rule provides a threshold for regulatory review, not the final determination of safety, for proposed activities. After determining that a proposed activity is safe and effective through appropriate engineering and technical evaluations, the 10 CFR 50.59 process is applied to determine if a license amendment is required prior to implementation. The process involves three basic steps: (1) applicability and screening to determine if a 10 CFR 50.59 evaluation is required; (2) an evaluation that applies the eight evaluation criteria of 10 CFR 50.59 (c)(2) to determine if a license amendment must be obtained from the NRC; and, (3) documentation and reporting to the NRC of activities implemented under 10 CFR 50.59.

A licensee shall obtain a license amendment pursuant to 10 CFR 50.90 prior to implementing a proposed change, test, or experiment if the change, test, or experiment would: result in more than a minimal increase in the frequency of occurrence of a previously evaluated accident; result in more than a minimal increase in the likelihood of occurrence of a malfunction of a structure, system, or component (SSC) important to safety; result in more than a minimal increase in the consequences of a previously evaluated accident; result in more than a minimal increase in the consequences of a malfunction of an SSC important to safety; create a possibility for an accident of a different type than any previously evaluated; create a possibility for a malfunction of an SSC important to safety with a different result than any previously evaluated; result in a design basis limit for a fission product barrier being exceeded or altered; or result in a departure from a method of evaluation used in establishing the design bases or in the safety analyses. According to § 50.90, whenever a holder of a license or construction permit desires to amend the license or permit, it must file an application for an amendment with the Commission, as specified in § 50.4, fully describing the changes desired, and following as far as applicable, the form prescribed for original applications. The NRC performs and documents a safety evaluation in these instances before it authorizes the change.

10 CFR 50.71. Paragraph (e) of 10 CFR 50.71 sets forth another process for making changes. This regulation requires licensees to update their final safety analysis reports periodically to incorporate the information and analyses that they submitted to the Commission or prepared pursuant to Commission requirements. Revisions to the updated final safety analysis reports are to include the effects of changes that occur in the vicinity of the plant, changes made in the facility or procedures described in the report, safety evaluations for approved license amendments and for changes made under 10 CFR 50.59, and safety analyses conducted at the request of the Commission to address new safety issues.

14.1.1.2 Regulatory Framework for the Restart of Browns Ferry Unit 1

This section describes the assessment and verification of safety for a plant that was restarted after being shutdown for a number of years.

As background, the Browns Ferry site, located near Decatur, Alabama, has three BWRs (General Electric, BWR-4, Mark-1 containment). All three units were shut down in 1985 to address management and regulatory issues. After resolving these issues, TVA successfully restarted Units 2 and 3 in the 1990s, but kept Unit 1 in a defueled layup condition. In May 2002, TVA decided to initiate a restart effort for Unit 1. The three Browns Ferry units are similar in design and licensing basis. TVA has implemented programs for Unit 1 that are similar to those used to restart Units 2 and 3, incorporated improvements and lessons learned and dedicated resources, including personnel with experience on restarting Units 2 and 3. The restart of Unit 1 differed from the restart of Units 2 and 3 in that TVA applied simultaneously for both a license renewal and an extended power uprate for Unit 1.

The regulatory framework for the restart of Browns Ferry Unit 1 consisted of two major elements-inspection and licensing activities. The NRC performed inspections in accordance with Inspection Manual Chapter 2509 and conducted the licensing activities consistent with the NRC's August 2003 regulatory framework letter discussed below.

TVA has submitted many varied licensing actions. The regulatory framework letter, provided a detailed listing of generic communications and other licensing actions requiring regulatory review and approval as wells as follow-up inspection. To facilitate communications with key stakeholders, the NRC held periodic public meetings at the site and developed a public outreach Web page which is similar to the Reactor Oversight Process Web page.

As part of its inspection program, the NRC staff reviewed TVA programs and plant activities related to the recovery of Unit 1. These activities included replacement, renovation, and removal of equipment and a review of plant programs, process and training of plant personnel. The NRC inspections of structural, electrical, mechanical and fire protection modifications resulted in satisfactory findings. Onsite monitoring and review determined that activities involving replacement, renovation, and removal of equipment were satisfactorily conducted so as to maintain adequate nuclear and radiological safety.

In addition, the NRC conducted an Operational Readiness Assessment Team inspection in April 2007 to assess management controls, implementation of site programs and personnel readiness to support safe restart and operation of Unit 1. The purpose of this inspection was to focus on the effectiveness of licensee management oversight, safety-significant activities, operator training and experience, corrective action programs, the maintenance program, operator response to annunciators and general plant conditions impacting safety, and the readiness to support three-unit operations. The inspection concluded that site programs, personnel and procedures were adequate for restart of Unit 1 and three unit power operations.

On May 15, 2007, the NRC authorized TVA to restart Browns Ferry Unit 1. The unit was restarted on May 22, 2007 and reached 100 percent power on June 8, 2007. TVA will conduct post restart testing, including the performance of two large transients tests, scheduled for completion within a few months.

After extensive reviews and inspections over the past several years, and completion of the regulatory framework issues, the NRC oversight for all cornerstones is now being conducted in accordance with the reactor oversight process. However, due to the lack of valid historical plant specific data for the remaining three cornerstones, additional reactor oversight process baseline inspections will be necessary until sufficient plant specific data becomes available.

14.1.2 License Renewal

This section explains license renewal. It discusses the governing documents and regulatory process, and recent experience, and provides relevant examples.

14.1.2.1 Governing Documents and Process

Background. The Atomic Energy Act and NRC regulations limit commercial power reactor licenses to 40 years but permit such licenses to be renewed. The original 40-year term was selected on the basis of economic and antitrust considerations, not technical limitations.

The NRC has established a license renewal process that can be completed in a reasonable period of time with clear requirements to ensure safe plant operation for up to an additional 20 years of plant life. The NRC's current schedule is to complete renewal reviews within 30 months of receipt of the application if a hearing is conducted, and within 22 months if not. The NRC completes nonstandard applications according to a schedule agreed upon with the applicant. The decision to seek license renewal rests entirely with nuclear power plant owners and typically is based on the plant's economic situation and whether it can meet NRC requirements.

Research has concluded that aging phenomena are readily manageable and do not pose technical issues that would preclude life extension for nuclear power plants. Studies have also found that facilities deal adequately with many aging effects during the initial license period, and credit should be given for these existing programs, particularly those under NRC's Maintenance Rule (10 CFR 50.65), which helps manage plant aging.

The license renewal process proceeds along two tracks—one for review of safety issues and another for environmental issues. An applicant must provide the NRC with an evaluation that addresses the technical aspects of plant aging and describes the ways it will manage those effects. It must also prepare an evaluation of the potential impact on the environment if the plant operates for up to another 20 years. The NRC reviews the application and verifies the safety evaluations through inspections.

Public participation is an important part of the license renewal process. Members of the public have opportunities to question how aging will be managed during the period of extended operation, and all information related to the review and approval of a renewal application is publicly available. Significant concerns may also be litigated in an adjudicatory hearing if any party who would be adversely affected requests a hearing.

10 CFR Part 54, "Requirements for Renewal of Operating Licenses for Nuclear Power Plants." Known as the License Renewal Rule, 10 CFR Part 54 establishes the technical and procedural requirements for renewing operating licenses. License renewal requirements for power reactors are based on two key principles:

(1)   When continued into the extended period of operation, the regulatory process, which assesses and verifies safety, is adequate to ensure that the licensing basis of all currently operating plants provides an acceptable level of safety. The possible exception is detrimental effects of aging on certain SSCs, and possibly a few other issues applying to safety only during the period of extended operation.

(2)   Each plant must maintain its licensing basis throughout the renewal term.

Guidance that applies to license renewal includes Regulatory Guide 1.188, "Standard Format and Content for Applications to Renew Nuclear Power Plant Operating Licenses," issued July 2001, to assist applicants in applying to renew a license and NUREG-1800, which guides the staff in reviewing applications. The Standard Review Plan incorporates by reference NUREG-1801, "Generic Aging Lessons Learned (GALL) Report," issued July 2001, which generically documents the basis for determining when existing programs are adequate and when they should be augmented for license renewal. As lessons are learned from the review of renewal applications or generic technical issues are resolved, the NRC issues improved guidance for interim use by applicants until the guidance is incorporated into the next formal update of the documents.

10 CFR Part 51, "Environmental Protection Regulations for Domestic Licensing and Related Regulatory Functions." The NRC's environmental protection regulation, 10 CFR Part 51, also applies to license renewal. This regulation was amended to facilitate the agency's environmental review process for license renewal. Specifically, the review requirements for 10 CFR Part 51 are founded on the conclusion that certain environmental issues can be resolved generically and need not be evaluated in each plant-specific application. These issues are described in NUREG-1437, "Generic Environmental Impact Statement for License Renewal of Nuclear Plants," issued May 1996. The NRC performs plant-specific reviews of the environmental impacts of license renewal to determine whether the impacts are so great that they should preclude license renewal as an option for energy-planning decisionmakers.

Supplement 1 to NRC Regulatory Guide 4.2, "Preparation of Supplemental Environmental Reports for Applications to Renew Nuclear Power Plant Operating Licenses," issued August 1991, provides guidance to applicants preparing environmental reports for license renewal. NUREG-1555, "Standard Review Plans for Environmental Reviews for Nuclear Power Plants, Supplement 1: Operating License Renewal," issued March 2000, guides the NRC staff's review of the environmental issues associated with a renewal application

14.1.2.2 Experience

The NRC issued the first renewed licenses for the Calvert Cliffs Nuclear Power Plant and the Oconee Nuclear Station in 2000. On the basis of industry statements, the NRC expects that essentially all remaining plants will apply for license renewal.

14.1.3 The United States and Periodic Safety Reviews

This report has retained the section on periodic safety reviews (PSRs) because of widespread interest in this topic at the Convention's third review meeting.

This section explains how the U.S. regulatory approach provides a continuum of assessment and review that ensures public health and safety throughout the period of plant operation. As discussed below, plant safety is improved by a combination of the ongoing NRC regulatory process, oversight of the current licensing basis, backfitting, broad-based evaluations, license renewal, and licensee initiatives that go beyond the regulations.

14.1.3.1 The NRC's Robust and Ongoing Regulatory Process and the Current Licensing Basis

Before issuing an operating license, the NRC comprehensively determines that the design, construction, and proposed operation of the nuclear power plant satisfy the NRC's requirements and reasonably ensure the adequate protection of the public health and safety. However, the licensing basis of a plant does not remain fixed for the 40-year term of the operating license. The licensing basis evolves throughout the term of the operating license because of the continuing regulatory activities of the NRC, as well as the activities of the licensee.

The NRC engages in many regulatory activities which, when considered together, constitute a process that provides ongoing assurance that the licensing bases of nuclear power plants provide an acceptable level of safety. This process includes inspections (both periodic regional inspections as well as daily oversight by the resident inspector), audits, investigations, evaluations of operating experience, regulatory research, and regulatory actions to resolve identified issues. The NRC's activities may result in changes to the licensing basis for nuclear power plants through promulgation of new or revised regulations, acceptance of licensee commitments to modify nuclear power plant designs and procedures, and the issuance of orders or confirmatory action letters. The agency also publishes the results of operating experience analysis, research, or other appropriate analyses through generic communication documents such as bulletins and generic letters. Licensee commitments in response to these documents also change the plant's licensing basis. In this way, the NRC's consideration of new information provides ongoing assurance that the licensing basis for the design and operation of all nuclear power plants provides an acceptable level of safety. This process continues for plants that receive a renewed license to operate for 20 years beyond the original operating license.

In addition to NRC-required changes in the licensing basis, a licensee may also voluntarily seek changes to the current licensing basis for its plant. However, these changes are subject to the NRC's formal regulatory controls on changes (such as 10 CFR 50.54; 10 CFR 50.59; 10 CFR 50.90; and 10 CFR 50.92, "Issuance of Amendment"). These regulatory controls ensure that licensee-initiated changes to the licensing basis for a plant are documented and that the licensee obtains NRC review and approval before implementing changes to the licensing basis that meet the review thresholds in 10 CFR 50.59. The licensee must report to the NRC any changes or modifications it makes to the licensing basis without prior NRC review at least every 2 years. Region-based NRC inspectors perform a sampling inspection of those changes in accordance with the Reactor Oversight Process to ensure that the licensee has properly characterized the changes or modifications.

14.1.3.2 The Backfitting Process: Timely Imposition of New Requirements

The NRC recognized the need to consider new requirements systematically rather than depending on the license renewal process or other regulatory processes to decide on plant upgrades. In the late 1970s and early 1980s, the NRC recognized the need for a process to determine when to address generic issues for all plants. As a result, the NRC developed the "backfitting" process and initiated the Committee To Review Generic Requirements (CRGR) to review staff-proposed backfits on licensees.

Also known as the Backfit Rule, 10 CFR 50.109 applies to both generic and plant-specific backfits for power reactors. It defines a "backfit" as any modification of or addition to plant systems, structures, components, procedures, organizations, design approvals, or manufacturing licenses that may result from the imposition of a new or amended rule or regulatory staff position. Except in the case of backfits that are imposed to bring a licensee back into compliance with its license or to ensure adequate protection of the public health and safety or common defense and security, the rule requires a cost-benefit backfit analysis. The NRC must determine though a backfit analysis that the proposed backfit will substantially increase the overall protection of the public health and safety (or common defense and security) and that the direct and indirect costs for the facility are justified in view of the increased protection.

Compliance and adequate protection backfits are justified differently. A documented evaluation is required, which provides the basis and states the objectives and purpose of the proposed backfit.

In 1988, the NRC issued an amended Backfit Rule, which clearly states that economic costs will not be considered in cases of ensuring, defining, or redefining adequate protection of the public health and safety, or in cases of ensuring compliance with NRC requirements or written licensee commitments.

Backfitting is expected to occur and is an inherent part of the regulatory process. However, it is permitted only after a formal, systematic review to ensure that changes are properly justified and suitably defined. The requirements of this process are intended to ensure order, discipline, and predictability and to optimize use of NRC staff and licensee resources.

The controls on generic backfitting include review by the CRGR, a committee of senior managers from various NRC offices. Established in 1981, this committee operates under a charter that specifically identifies the documents to be reviewed and the analyses, justifications, and findings to be provided. Its objectives include eliminating unnecessary burdens on licensees, reducing radiation exposure to workers while implementing requirements, and optimizing use of NRC and licensee resources to ensure safe operation. Thus, the CRGR charter is a key implementing procedure for generic backfitting, although the primary responsibility for proper backfit considerations belongs to the initiating organization.

14.1.3.3 The NRC's Extensive Experience with Broad-Based Evaluations

In the mid-1970s, the NRC recognized the importance of assessing the adequacy of the design and operation of currently licensed nuclear power plants, understanding the safety significance of deviations from applicable current safety standards that may have been approved after those plants were licensed, and providing the capability to make integrated and balanced decisions about the need for backfit modifications at those plants.

Consequently, in 1977, the NRC initiated the Systematic Evaluation Program (SEP). From a list of approximately 800 potential issues and topics related to nuclear safety, the SEP found that the regulatory requirements for 137 issues had changed sufficiently to warrant evaluation. The staff then compared the designs of 10 of the older plants to the licensing criteria delineated in the then recently issued Standard Review Plan.[3] After further review, the staff determined that 27 issues required some corrective action at one or more plants and resolution of those issues could lead to safety improvements at other operating plants built at about the same time. These 27 issues became known as the 27 "SEP lessons learned."

In 1984, NRC staff presented the 27 SEP lessons learned to the Commission as part of a proposal for an Integrated Safety Assessment Program (ISAP). The staff developed this program to review safety issues for a specific plant in an integrated manner instead of continuing the SEP at other older operating reactors. In November 1984, the Commission published the "Commission Policy Statement on the Systematic Evaluation of Operating Nuclear Power Reactors." In this policy statement, the Commission articulated its view that issues relating to the safety of operating nuclear power plants can be more effectively and efficiently implemented in an integrated, plant-specific review. For the first time, the Commission discussed probabilistic safety analysis as a method to obtain consistent and comparable results which could be used to enhance a safety assessment. The SEP process was transformed into the ISAP pilot program.

In May 1985, the NRC initiated the ISAP pilot program at two plants, Millstone Unit 1 and Haddam Neck (Connecticut Yankee). The ISAP pilot program identified some benefits; however, the Commission deferred extending it beyond the pilot phase until the staff provided an integrated package of options that clarified the relationship between the proposed follow-on program to the ISAP pilot (ISAP II) and the newly proposed individual plant examination (IPE) process.

The Commission determined that, since the ISAP II program would be voluntary and the IPE program, through the NRC's generic letter process, would require a licensee response, it should give priority to the IPE program. Many of the same benefits that might have been derived through the proposed ISAP II were derived instead through the IPE (e.g., probabilistic safety analysis) process.

In the late 1980s and throughout the 1990s, the NRC continued its efforts to strengthen its regulatory infrastructure and ensure continued safe operation of commercial nuclear power plants through inspection, broad-based assessment, and where appropriate, establishment of new generic requirements. For example, the Commission determined that licensees should assess the accessibility and adequacy of their design-basis information and determine whether their plants needed a design-basis reconstitution program. The Commission expressed its expectations in "Availability and Adequacy of Design Bases Information at Nuclear Power Plants;

---

[3]   Standard Review Plans help ensure the quality and uniformity of staff reviews and provide a well-defined base from which to evaluate a licensee or applicant submittal. The Standard Review Plans are also intended to make information about regulatory matters widely available, to enhance communication with interested members of the public and the nuclear power industry, and to improve the understanding of the staff review process.

Policy Statement," in the *Federal Register* of August 10, 1992. The Commission also expanded the IPE program to consider external events and, recognizing the relationship between maintenance, equipment reliability, plant risk, and safety, the Commission promulgated the Maintenance Rule.

14.1.3.4 *License Renewal Confirms Safety of Plants*

As late as 1991, some plants had not definitively resolved the 27 SEP lessons learned. As the staff considered a process to renew the operating licenses for the operating nuclear power plants, it assessed the best way to address these 27 issues.

Of the 27 issues, four had been completely resolved for all plants. One other issue was of such low safety significance that it required no additional action. The staff determined that none of the remaining 22 issues required immediate action to protect public health and safety. The staff placed these 22 issues into the established regulatory process for determining the safety significance of generic issues.[4]

In developing the License Renewal Rule, the Commission concluded that issues material to the renewal of a nuclear power plant operating license are limited to those issues that the Commission determines are uniquely relevant to protecting the public health and safety and preserving common defense and security during the period of extended operation. Other issues would, by definition, be relevant to the safety and security of the public during current plant operation. Given the Commission's ongoing obligation to oversee the safety and security of operating reactors, the existing regulatory process within the present 40-year license term would address issues related to current plant operation rather than deferring the issues until the time of license renewal. (See Section 14.1.2 for a description of license renewal.) To add to that discussion, license renewal applicants are required to complete an integrated plant assessment (IPA)[5] and evaluate time-limited aging analyses.

14.1.3.5 *Risk-Informed Regulation and the Reactor Oversight Process*

The NRC is actively increasing the use of risk insights and information in its regulatory decision-making. In the reactor area, risk-informed activities occur in the five broad categories of (1) applicable regulations, (2) licensing process, (3) revised oversight process, (4) regulatory

---

[4] A generic issue is a regulatory matter that is not sufficiently addressed by existing regulations, guidance, or programs. The NRC has identified by its systematic assessment of plant operation certain issues that seem prevalent among plants. The NRC documents and tracks resolution of these "generic safety issues." The generic safety issue program provides for (1) identifying generic issues, (2) assigning them priorities, (3) developing detailed action plans for their resolution, (4) overseeing progress in their resolution by senior managers, and (5) informing the public of the status of progress in resolution. The resolution of these issues may involve new or revised rules, new or revised guidance, or revised interpretation of rules or guidance that affect nuclear power plant licensees or nuclear material certificate holders. Congress requires that the NRC maintain this program.

[5] An IPA identifies and lists structures and components subject to an aging management review. These include "passive" structures and components that perform their intended function without moving parts or without a change in configuration or properties. Examples of these are components such as the reactor vessel, the steam generators, piping, component supports, and seismic Category I structures. To be in scope, the item must also be "long-lived" to be considered during the license renewal process. Long-lived means the item is not subject to replacement based on a qualified life or specified time period.

guidance, and (5) risk analysis tools, methods, and data. Activities within these categories include revisions to technical requirements in the regulations; risk-informed technical specifications; a new framework for inspection, assessment, and enforcement actions; guidance on risk-informed inservice inspections; and improved standardized plant analysis risk models.

In 2000, the NRC implemented a revised Reactor Oversight Process (see discussion under Article 6) using risk insights and lessons learned from more than 40 years of regulating nuclear power plants. The previous oversight process evolved during a period when the nuclear power industry was less mature and there was much less operational experience on which to base rules and regulations. Very conservative judgments governed the rules and regulations. Significant plant operating events occurred with some frequency, and the oversight process tended to be reactive and prescriptive, closely observing plant performance for adherence to the regulations and responding to operational problems as they occurred.

After nearly four decades of operational experience and generally steady improvements in plant performance, the Reactor Oversight Process now focuses more of the agency's resources on the relatively small number of plants that evidence performance problems. The Reactor Oversight Process is more effective in correcting performance or equipment problems today because the agency's response to problems is more timely and predictable.

The Reactor Oversight Process makes greater use of objective performance indicators. Together, the indicators and inspection findings provide the information needed to support reviews of plant performance, which are conducted quarterly. In addition, the Reactor Oversight Process features expanded semiannual reviews, which include inspection planning and a performance report (all of which are posted on the NRC's public Web site).

14.1.3.6 Licensee Responsibilities for Safety: Regulations and Initiatives Beyond Regulations

As in many countries, U.S. nuclear power plant licensees are responsible for the safety of their facilities. This responsibility is embedded in their license and in the NRC's regulatory infrastructure. Under the regulatory umbrella, licensees routinely assess new technologies, off-normal conditions, operating experience, and industry trends to make informed decisions about safety enhancements to their facilities.

The NRC does not specifically mandate some of these reviews. Rather, they are self-imposed initiatives over and above regulations, motivated by the licensees' self-described pursuit of excellence and by the recognition that, in the U.S. free-market competitive energy industry, safety and economics are directly linked. Licensees have, for example, voluntarily replaced analog instrumentation and control systems with digital systems, upgraded their plants to increase production of electricity, and managed their plants to performance levels above the NRC's performance indicator thresholds.

Under the U.S. regulatory structure, Appendix B to 10 CFR Part 50 requires that all nuclear power plant licensees maintain a QA program. QA comprises all those planned and systematic actions necessary to provide adequate confidence that an SSC will perform satisfactorily in service. QA includes quality control, which comprises those QA actions related to the physical characteristics of a material, structure, component, or system that provide a means to control quality to predetermined requirements.

Licensees carry out a comprehensive system of planned and periodic audits to verify compliance with all aspects of the QA program and to determine the effectiveness of the program. Appropriately trained personnel who do not have direct responsibilities in the areas being audited perform these audits in accordance with written procedures or checklists. Audit results are documented and reviewed by management with responsibility in the area audited, and appropriate followup is initiated.

14.1.3.7 Summary

IAEA and the Western European Nuclear Regulators' Association (WENRA) have developed guidance[6] and objectives for conducting PSRs that have much in common. Specifically, consistent with the guidance of both organizations, PSRs are comprehensive assessments with the following purposes:

- to determine, at the time of the review, whether the plant complies with its licensing basis

- to identify the extent to which the current licensing basis remains valid, in part by determining the extent to which the plant meets current safety standards and practices

- to provide a basis for implementing appropriate safety improvements, corrective actions, or process improvements

- to provide confidence that the plant can continue to be operated safely

For the reasons discussed above and summarized below, the shared objectives associated with IAEA and WENRA PSR guidance are substantively accomplished in the United States on an ongoing basis.

First, the NRC's regulatory process provides a robust foundation for ongoing assessments, evaluations, and when appropriate, imposition of new requirements. The NRC and the U.S. nuclear industry consider new information, in a more risk-informed manner, as it becomes available; adjust the regulatory oversight and plant safety priority, respectively; and provide ongoing assurance that the licensing basis for the design and operation of all nuclear power plants provides an acceptable level of safety.

Second, the NRC and the U.S. nuclear industry have a 30-year history of implementing broad-based plant assessments. The regulatory history of implementing broad-based assessments is a direct result of an adaptive, probing, and independent regulatory process. These assessments have included the SEP, the ISAP, the IPE, and the reactor license renewal process and provide additional confidence that plant safety continues to be the highest priority and that the NRC and industry continue to pursue enhancements that improve safety. The time line in Figure 4 demonstrates that, over a period of almost 25 years, broad-based NRC assessments and regulatory initiatives have provided a continuum of assessment, improvement, and oversight, which ensures that licensed plants continue to operate safely.

---

[6]   IAEA guidance appears in Safety Standards Series No. NS-G-2.10, "Periodic Safety Review of Nuclear Power Plants Safety Guide,"issued in 2003. WENRA guidance appears in "Pilot Study on Harmonization of Reactor Safety in WENRA Countries," WENRA Working Group on Reactor Harmonization, March 2003.

The NRC's approach to continuing to ensure plant safety differs from the historically deterministic focus of PSRs. The transition to a more risk-informed regulatory framework and the Reactor Oversight Process provides an ongoing approach and basis for implementing appropriate safety improvements, corrective actions, or process improvements and provides confidence that the plant can continue to be operated safely. The NRC's more risk-informed approach helps ensure that resources are optimally focused on those issues most important to safety.

Finally, U.S. licensees establish performance expectations above the thresholds required by the NRC. These self-imposed expectations and initiatives—over and above the regulations—result from the licensee's self-described motivation to pursue excellence and by the recognition that, in the free-market competitive industry in the United States, safety and economics are directly linked.

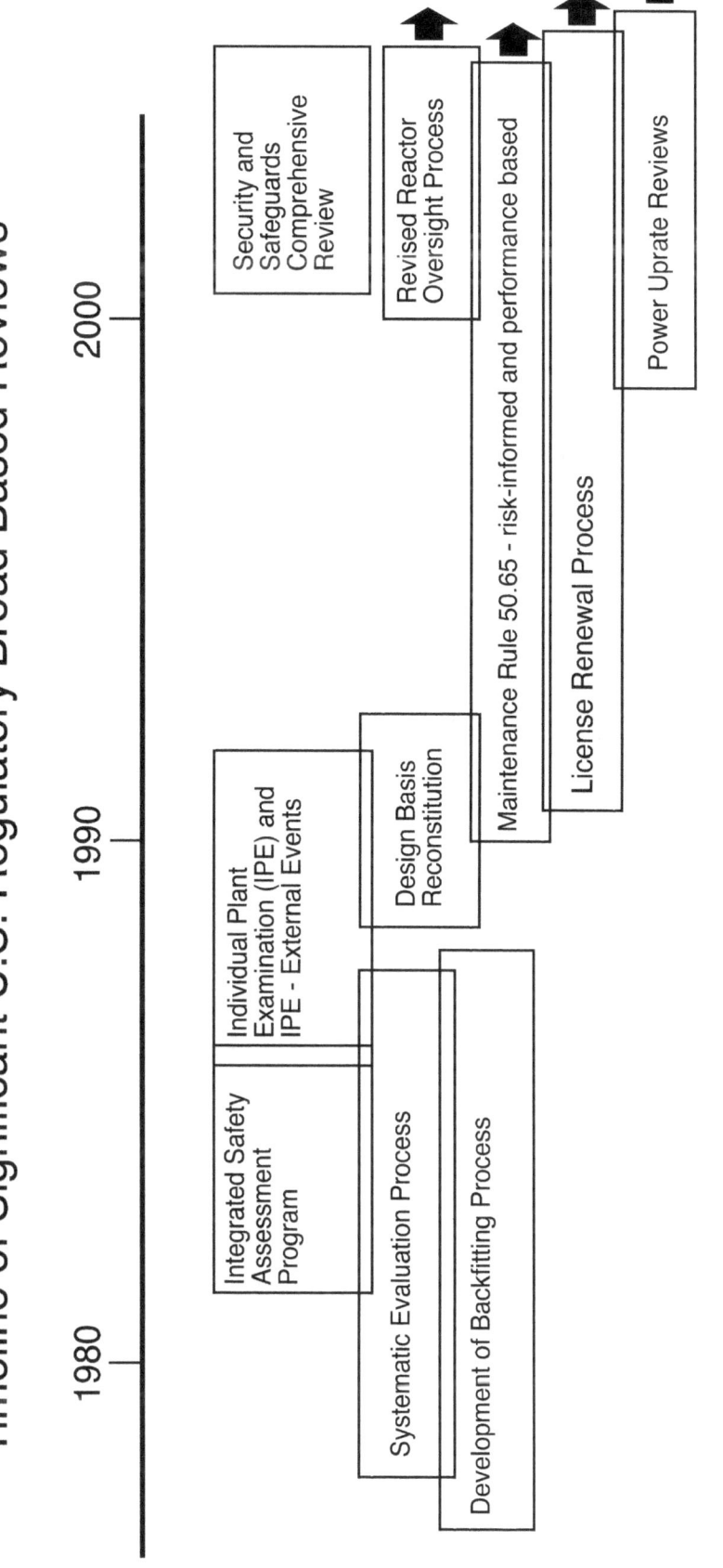

Timeline of Significant U.S. Regulatory Broad Based Reviews

101

14.2 Verification by Analysis, Surveillance, Testing, and Inspection

Licensees are required to verify that they are operating their nuclear installations in accordance with the plant-specific design and requirements. The technical specifications (for surveillance) and national consensus codes (for testing and periodic inspections) contain the requirements specifying verification.

In 10 CFR 50.55a, "Codes and Standards," the requirements appear for applying industry codes and standards to nuclear power reactors during design, construction, and operation. This section states, "Systems and components of boiling and pressurized water-cooled nuclear power reactors must meet the requirements of the ASME Boiler and Pressure Vessel Code specified in paragraphs (b) through (g) of this section." In addition, 10 CFR 50.55a provides for alternatives to the ASME Code when authorized by the NRC.

Through analysis, surveillance, testing, and inspection, the NRC verifies that the physical state and operation of nuclear installations continue to be in accordance with the designs, applicable national safety requirements, and operational limits and conditions. As previously discussed in Article 6, the NRC's Reactor Oversight Process includes inspections to verify that licensees are fulfilling their obligations to conduct such surveillances and testing and take corrective action. The NRC staff updates, revises, and improves existing regulatory programs in light of operating experience and significant new safety information. Article 19 discusses these activities.

The Commission may also require under 10 CFR 50.54(f) "Conditions of licenses" that licensees submit written statements to enable the Commission to determine whether or not the license should be modified, suspended, or revoked.

ARTICLE 15. RADIATION PROTECTION

Each Contracting Party shall take the appropriate steps to ensure that, in all operational states, the radiation exposure to the workers and to the public caused by a nuclear installation shall be kept as low as is reasonably achievable, and that no individual shall be exposed to radiation doses that exceed the prescribed national dose limits.

This section summarizes the authorities and principles of radiation protection, which include the regulatory framework, regulations, and radiation protection programs for controlling radiation exposure for occupational workers and members of the public. Article 17 addresses radiological assessments that apply to licensing and to facility changes.

The changes in this section are an updating of doses and an expanded discussion of Appendix I, "Numerical Guides for Design Objectives and Limiting Conditions for Operation to Meet the Criterion 'As Low as Is Reasonably Achievable' for Radioactive Material in Light-Water-Cooled Nuclear Power Reactor Effluents," to 10 CFR Part 50; work on the International Commission on Radiological Protection (ICRP) recommendations; and ground water contamination.

15.1 Authorities and Principles

Generally, U.S. radiation control measures are founded on radiological risk assessments by the United Nations Scientific Committee on the Effects of Atomic Radiation and the U.S. National Academy of Sciences Committee on the Biological Effects of Ionizing Radiation. The risk management recommendations promulgated by the ICRP and the National Council on Radiation Protection and Measurements (NCRP) reflect these assessments. On the basis of these assessments and recommendations, the EPA develops "generally applicable radiation standards" for use by the other Federal agencies, including the NRC. Considering these recommendations and standards, the responsible agencies, such as the NRC, then establish regulations.

The principles that are the basis of the U.S. radiation protection programs are generally consistent with the principles espoused by the ICRP. That is to say, (1) it is known that large doses of ionizing radiation can be deleterious to human health, and (2) it is considered prudent to assume that small doses may also be harmful, with the probability of a deleterious effect being proportional to the dose. The ICRP-recommended protection principles of "limitation," "justification," and "optimization" are acknowledged but are proving difficult to implement.

Of these principles, "limitation" is the most practicable. The regulations establish dose limits, and these limits cannot be exceeded without violating the regulations. There is a lengthy history of the doses being kept within the limits for workers (NUREG-0713, "Occupational Radiation Exposure at Commercial Nuclear Power Reactors and Other Facilities," Volume 24, issued October 2003) and members of the public living near nuclear power plants (NUREG/CR-2850, "Dose Commitments Due to Radioactive Releases from Nuclear Power Plant Sites in 1992," Volume 14, issued March 1996).

"Justification," the recommendation that any activity involving radiation exposure should be shown to be beneficial before the activity is undertaken, has proved on occasion difficult to demonstrate. The risks or benefits of a new application of radioactive material can seldom be determined in advance with complete accuracy. The "justification" activities in the United States are generally limited to the licensing process. In general, the NRC will reject an application to use or produce radioactive materials if it determines that the application is frivolous (i.e., that the overall benefit to society is outweighed by the risk of the radiation exposure associated with the activity). For some large applications, such as the generation of electricity from nuclear power, national policy establishes the justification. Since the National Energy Policy favors nuclear power (i.e., the net benefit for the United States is deemed to be positive), the licensing process under 10 CFR Part 50 does not specifically address the justification for licensing a nuclear power plant.

Rather than "optimization," the United States has used the concept of ALARA, although the two principles are consistent. As a guiding principle, ALARA (with varying terminology) dates back to 1939 (at least in the United States) and is defined in the regulations for occupational workers and members of the public.

For decades, 10 CFR Part 20 has addressed the ALARA criterion for occupational radiation exposure but more as an admonition than as a requirement. In 1994, the regulation was changed to require that all licensees develop, document, and carry out an ALARA program. The NRC would judge compliance with this requirement on the basis of a licensee's capability to track and, if necessary, reduce exposures, and not on whether exposures and doses represented an absolute minimum or whether the licensee had used all possible methods to reduce exposures.

For control of radiation exposure to members of the public, the NRC modified 10 CFR Part 50 by adding Appendix I. Issued in 1975, this appendix requires that radioactive releases from nuclear power plants be kept ALARA. This requirement led to the establishment of numerical objectives (i.e., 0.00005 sievert (Sv) (0.005 rem) in a year to the most highly exposed individual). Similar EPA requirements for other facilities soon followed this NRC requirement. It is not clear that these requirements satisfy the intent of the ICRP, but they are sufficient to keep public doses well below the local variation in doses from natural sources.

Although U.S. regulations are generally consistent with ICRP recommendations, to date, certain constraints have limited the extent to which the U.S. regulations coincide with those of the ICRP. One important constraint has been the desire for regulatory stability. Revising the regulations to incorporate every new ICRP position would impose a serious burden on the licensees without a commensurate benefit. Furthermore, for nuclear power reactors, new requirements are constrained by the Backfit Rule (10 CFR 50.109), which essentially requires that any increase in regulatory requirements be justified by a commensurate improvement in safety. Consequently, U.S. regulations were founded on older (rather than the most recent) recommendations of the ICRP. Nevertheless, the Commission has directed NRC staff to work closely with the ICRP and other national and international organizations to assist in developing the 2007 ICRP recommendations. The NRC may revise its regulations, in whole or in part, depending on the nature of these recommendations.

15.2 Regulatory Framework

Requirements for radiation protection were developed to implement laws passed by Congress. These laws are the Atomic Energy Act, the Energy Reorganization Act, and the Uranium Mill Tailings Radiation Control Act of 1978.

NRC regulations establish the primary direct controls over licensees. Various documents provide additional guidance and clarification. Specifically, these documents include regulatory guides, topical staff and contractor reports (NUREG series), generic letters, technical specifications, and license conditions. These documents are supported by international standards, consensus national standards, and authoritative recommendations (such as those of the ICRP and NCRP). However, these supporting documents have no official status unless they are referenced in or adopted by a regulation or documents providing regulatory guidance, such as regulatory guides or Standard Review Plans. Of particular importance are NUREG-0800, which guides the staff in reviewing safety analysis reports, and Regulatory Guide 1.70, "Standard Format and Content of Safety Analysis Reports for Nuclear Power Plants," Revision 3, issued in November 1978, which guides the applicant in writing safety analyses. Chapter 11 of NUREG-0800 addresses the control of radioactive effluents. Chapter 12 addresses radiation protection. Chapter 15 details how to calculate offsite and control room operator doses for design-basis accidents. Paragraph (g) of 10 CFR 50.34 requires the evaluation of the facility against the Standard Review Plan.

As discussed under Article 6, the Reactor Oversight Process has cornerstones for radiation safety. The cornerstone Public Radiation Safety focuses on the effectiveness of the plant's programs to meet applicable Federal limits involving the exposure, or potential exposure, of members of the public to radiation and to ensure that the effluent releases from the plant are ALARA. The cornerstone Occupational Radiation Safety focuses on the effectiveness of the plant's program(s) in maintaining the worker dose within the regulatory limits and providing occupational exposures that are ALARA.

15.3 Regulations

The regulations that apply to radiation protection are 10 CFR Part 20 and 10 CFR Part 50.

10 CFR Part 20. This part of the NRC regulations establishes requirements for radiation protection for all NRC licensees. Specific requirements for specific operations and specific kinds of licenses supplement the requirements in 10 CFR Part 20. In particular, these supplementary requirements include 10 CFR Part 30, "Rules of General Applicability to Domestic Licensing of Byproduct Material"; 10 CFR Part 34, "Licenses for Industrial Radiography and Radiation Safety Requirements for Industrial Radiographic Operations"; 10 CFR Part 35, "Medical Use of Byproduct Material"; 10 CFR Part 39, "Licenses and Radiation Safety Requirements for Well Logging"; 10 CFR Part 40, "Domestic Licensing of Source Material"; 10 CFR Part 50; 10 CFR Part 70, "Domestic Licensing of Special Nuclear Material"; 10 CFR Part 71, "Packaging and Transportation of Radioactive Material; and 10 CFR Part 72, "Licensing Requirements for the Independent Storage of Spent Nuclear Fuel, High-Level Radioactive Waste, and Reactor-Related Greater than Class C Waste."

The most recent major revision of 10 CFR Part 20, issued in 1991, adopted the recommendations, quantities, and models recommended in ICRP Publication 26, "Recommendations of the International Commission on Radiological Protection (Adopted January 17, 1977)," issued in 1991, and ICRP Publication 30, "Limits of Intakes of Radionuclides by Workers," dated 1978–1982, as well as some recommendations from NCRP Report No. 91, "Recommendations on Limits for Exposure to Ionizing Radiation," issued June 1987. The regulations in 10 CFR Part 20 provide relatively comprehensive coverage of general requirements for radiation protection and 10 CFR Part 20 itself is divided into subparts, with each subpart addressing a specific area of radiation protection, such as occupational and public dose limits, positing, surveys, monitoring, waste disposal, and reporting.

The details of the requirements in 10 CFR Part 20 are not entirely consistent with international standards such as IAEA's Basic Safety Standards. The main areas of difference include use of the effective dose equivalent in 10 CFR Part 20 versus use of the effective dose in the Basic Safety Standards; an annual occupational dose limit on the effective dose equivalent of 0.05 Sv in 10 CFR Part 20 versus 0.02 Sv in the Basic Safety Standards; and use of the ICRP-30 biokinetic models in 10 CFR Part 20 versus the more recent models used in the Basic Safety Standards. The NRC is planning to revise its regulations in the near future to bring them closer to international standards. However, in the interim, NRC licensees are permitted to use the effective dose in place of the effective dose equivalent and to use the more recent internal dosimetry models in place of those recommended in ICRP-30, with prior NRC approval. In addition, many licensees and agencies have administrative dose limits that are similar to, or lower than, those in the Basic Safety Standards, and most other licensees operate at occupational doses far below those limits and standards, and therefore, are considered ALARA. In some cases, the occupational doses do exceed 0.2 Sv per year, but these are a very small fraction of the total, and efforts are continuing to reduce these doses to lower levels. In the interim, and until NRC's regulations are brought into closer formal conformance with international standards, the current 10 CFR Part 20 provides a level of radiation protection that in almost all situations is comparable to that provided by international standards.

10 CFR Part 50. This is the principal regulation that addresses the safety of nuclear power plants. However, only a small part directly addresses radiation protection. (The revised dose criteria for design-basis accidents appear in 10 CFR 50.34(a)(1)(ii)(D) for future licensing actions after implementation of the revised rule in 1997. The dose criteria for siting and determining the exclusion area low population zone and population center distance for nuclear power reactors are stated in 10 CFR 100.11(a).) Even so, the sections of 10 CFR Part 50 that do affect radiation protection are significant. Of particular importance are 10 CFR 50.34a, "Design Objectives for Equipment to Control Releases of Radioactive Material in Effluents—Nuclear Power Reactors," and Appendix I to 10 CFR Part 50 and 10 CFR 50.34(g), which requires NRC review of the in-plant radiation protection program.

15.4 Radiation Protection Activities

Radiation protection activities apply to occupational workers and to members of the public.

15.4.1 Control of Radiation Exposure of Occupational Workers

In addition to focusing on personnel qualifications for licensing, the NRC's oversight and regulation of the radiation protection programs ensure that the safety analysis report and radiation protection plan properly address each item in 10 CFR Part 20, as well as the "Instruction to Workers" provisions of 10 CFR Part 19, "Notices, Instructions, and Reports to Workers: Inspection and Investigations," and the provisions of the relevant regulatory guides, such as Regulatory Guide 1.8, "Personnel Selection and Training," issued March 1971, and Regulatory Guide 8.8, "Information Relevant to Ensuring that Occupational Radiation Exposures at Nuclear Power Stations Will Be As Low As Is Reasonably Achievable," Revision 3, issued June 1978.

Once the NRC issues a license, it maintains an active regulatory program, which includes routine monitoring of licensee and regional reports to alert NRC staff of potential problems in radiation safety. Significant health physics problems can trigger significant reactive regional inspections or a generic communication to the industry.

NRC staff has been collecting the annual occupational exposure data for light-water reactors since 1969. The doses are strongly influenced by the amount and kind of maintenance performed, so the individual plant collective doses fluctuate from year to year. Still, clear trends are evident. Using the average collective dose per reactor as the reference statistic, one can conclude that the doses were almost randomly variable before the accident at TMI Unit 2. Thereafter, the doses increased as a result of the extensive modifications required of all nuclear power plants in response to new NRC requirements. The average collective dose reached a peak of 7.91 person-Sv (791 person-rem) per reactor in 1980. Since then, doses have declined almost steadily to the current level of slightly above 1 person-Sv (100 person-rem) per reactor, where they have remained for the past 8 years (1998–2005, the last year for which the data have been compiled). The 2004 average collective dose value of 1.0 person-Sv (100 person-rem) per reactor was the lowest average collective dose recorded since data collection began in 1969. Although the average doses for both PWRs and BWRs have been steadily declining, the average BWR dose has exceeded the average PWR dose since 1974. Over the past 5 years, the average BWR dose has exceeded the average PWR dose by roughly 90 percent (in part, because of the higher average dose rates and larger work force at BWRs). In 2005, the 78,127 workers at nuclear plants received 115 person-Sv (11,456 person-rem) for an average of 0.0015 Sv (0.15 rem) per worker. This represents an 84-percent drop in average worker dose from the 1973 value of 0.0095 Sv (0.95 rem) per worker.

15.4.2 Control of Radiation Exposure of Members of the Public

The regulations in 10 CFR 50.34a and Appendix I to 10 CFR Part 50 define the ALARA plant objectives for effluents. Appendix I also specifies effluent monitoring, environmental monitoring, investigations, land-use censuses, and reporting. Section IV.B of Appendix I to 10 CFR Part 50 requires the licensee to establish an appropriate surveillance and monitoring program that will:

> 1. Provide data on quantities of radioactive material released in liquid and gaseous effluents...;
>
> 2. Provide data on measurable levels of radiation and radioactive materials in the environment to evaluate the relationship between quantities of radioactive material released in effluents and resultant radiation doses to individuals from principal pathways of exposure; and
>
> 3. Identify changes in the use of unrestricted areas (e.g., for agricultural purposes) to permit modifications in monitoring programs for evaluating doses to individuals from principal pathways of exposure.

Appendix I requirements are supplemented by 10 CFR Part 20.1501, "General," which requires, in part, that a licensee perform surveys to evaluate potential radiological hazards and to demonstrate compliance with the public dose limits in 10 CFR 20.1301 and 10 CFR 20.1302.

Therefore, a licensee is responsible for performing radiation surveys at its facility to look for radioactive materials that have the potential to affect workers and members of the public. Potential survey sites can include areas that have been previously impacted by licensed radioactive material, as well as areas that may be impacted by licensed radioactive material in the future. For onsite spills and leaks that may contain licensed radioactive material, 10 CFR 20.1501 requires a licensee to conduct appropriate radiation surveys and monitoring to determine the radiological hazard (i.e., dose assessment) to workers and to determine if there is a viable pathway to the unrestricted area, which could result in a potential radiological hazard to members of the public. The surveys and monitoring can continue over a period of time or become an ongoing monitoring program so that the licensee can adequately characterize the extent and source of the contamination from the spills or leak.

In the past three years, there have been several discoveries of radioactive ground water contamination at nuclear power facilities located throughout the United States. Investigation has determined that most of the contamination resulted from undetected leakage from facility SSCs that contained or transported radioactive liquids. All unmonitored releases resulted in varying levels of onsite tritium ground water contamination, with one facility detecting low levels of tritium (below EPA drinking water standards) in offsite residential drinking wells. Current data show no immediate public health impact and a very low probability that there will be an impact in the future.

The NRC has responded to reports of ground water contamination by conducting inspections and assessing the safety significance of these events, in addition to evaluating licensee performance in identifying and taking corrective actions. The NRC has also issued Information Notices (IN 2004-05, "Spent Fuel Pool Leakage to Onsite Groundwater," dated March 3, 2004, and IN 2006-13, "Ground-Water Contamination Due to Undetected Leakage of Radioactive Water," dated July 10, 2006) describing unmonitored and unplanned leakage at several nuclear power stations.

Both the NRC and industry have worked to resolve the technical and programmatic issues leading to the ground water contamination events. In March 2006, the NRC Executive Director for Operations established a Liquid Radioactive Release Lessons Learned Task Force to assess lessons learned related to the unmonitored release of radioactive liquid to the environment at power reactor sites and to recommend possible agency actions in this area. The task force completed its assessment and issued its report on September 1, 2006. The most significant conclusion reached by the task force was that these events had no public health impact. However, because of the high level of public concern and the potential for contaminated ground water to migrate off site undetected, the task force made several recommendations to the NRC. These generally addressed enhanced regulations or regulatory guidance for unplanned, unmonitored releases and additional reviews in the areas of decommissioning funding and license renewal. The staff is currently evaluating all recommendations for implementation.

In parallel with the NRC efforts, the nuclear industry also responded to the ground water contamination events. The Nuclear Energy Institute has developed a voluntary Groundwater Protection Initiative that licensees have endorsed unanimously. The Groundwater Protection Initiative required each participating nuclear plant to have a plan in place by July 2006 that established several short-term actions, such as developing an enhanced communication protocol to ensure notification of State and local officials of less significant unmonitored release events. The industry initiative also required several long-term actions to improve leak detection monitoring capability and improve understanding of site hydrology and geology.

110

ARTICLE 16. EMERGENCY PREPAREDNESS

1.   Each Contracting Party shall take the appropriate steps to ensure that there are onsite and offsite emergency plans that are routinely tested for nuclear installations, and cover the activities to be carried out in the event of an emergency.

   For any new nuclear installation, such plans shall be prepared and tested before [the installation] commences operation above a low power level agreed [to] by the regulatory body.

2.   Each Contracting Party shall take appropriate steps to ensure that, insofar as they are likely to be affected by a radiological emergency, its own population and the competent authorities of the States in the vicinity of the nuclear installation are provided with appropriate information for emergency planning and response.

3.   Contracting Parties that do not have a nuclear installation on their territory, insofar as they are likely to be affected in the event of a radiological emergency at a nuclear installation in the vicinity, shall take appropriate steps for the preparation and testing of emergency plans for their territory that cover the activities to be carried out in the event of such an emergency.

This section discusses (1) emergency planning and emergency planning zones, (2) offsite emergency planning and preparedness, (3) emergency classification system and action levels, (4) recommendations for protection in severe accidents, (5) inspection practices and regulatory oversight, (6) response to an emergency, and (7) international arrangements.

This section was revised to describe the fundamental changes in response to national emergencies as a result of the terrorist events of September 11, 2001, as well as the response to Hurricane Katrina in August 2005.

16.1 Background

The NRC's responsibilities for radiological emergency preparedness stem from NRC licensing functions under the Atomic Energy Act and the Energy Reorganization Act. Both statutes specifically authorize the agency to promulgate regulations that it deems necessary to fulfill its responsibilities under the acts. Following the accident at TMI Unit 2 in March 1979, the regulations were amended to require significant changes in emergency planning and preparedness for U.S. commercial nuclear power plants. The NRC's emergency planning regulations are now an important part of the regulatory framework for protecting public health and safety and have been adopted as an added conservatism in the NRC's defense-in-depth safety philosophy of multiple-barrier containment and redundant safety systems. Before a full-power operating license can be issued, NRC regulations require a finding that there is reasonable assurance that adequate measures to protect public health and safety can and will be taken in a radiological emergency (10 CFR 50.47(a)).

Emergency planning in the United States recognizes that a spectrum of accidents could exceed the design-basis accidents that nuclear plants are required to accommodate without significant public health and safety impacts. For design-basis accidents, the small releases that might occur would not likely require responses such as evacuating or sheltering the general public. These actions become important only in considering accidents that are much less probable than design-basis accidents. NUREG-0396, "Planning Basis for the Development of State and Local Government Radiological Emergency Response Plans in Support of Light-Water Nuclear Power Plants," issued December 1978, and NUREG-0654/FEMA-REP-1 (NUREG-0654), "Criteria for Preparation and Evaluation of Radiological Emergency Response Plans and Preparedness in Support of Nuclear Power Plants," Revision 1, issued November 1980, describe the emergency planning basis.

16.2 Offsite Emergency Planning and Preparedness

The accident at TMI Unit 2 revealed that much better coordination and more comprehensive emergency plans and procedures were needed if the NRC and the public were to have confidence in the readiness of onsite and offsite emergency response organizations to respond to a nuclear emergency. Participation by State and local governments in emergency planning for nuclear power plants in the United States was, and still remains, largely voluntary. Before the accident at TMI 2, there had been no clear obligation for the State and local governments to develop emergency plans for radiological accidents, and the Federal role was one of assistance and guidance. After the accident, the NRC amended its emergency planning regulations to require, as a condition of licensing, that each applicant and licensee submit the radiological emergency response plans of State and local governments that are within the plume exposure zone, as well as the plans of State governments within the ingestion pathway zone (10 CFR 50.33(g) and 50.54(s)).

In December 1979, the President directed FEMA to take the lead in ensuring the development of acceptable State and local offsite emergency plans and activities for nuclear power facilities. The NRC and FEMA regulations and a memorandum of understanding between the two agencies, dated September 14, 1993, subsequently codified the role and responsibilities of DHS/FEMA.

DHS/FEMA provides its findings regarding the acceptability of the offsite emergency plans to the NRC, which has the ultimate responsibility for determining the overall acceptability of radiological emergency plans and preparedness for a nuclear power reactor. The NRC will not issue a license to operate a nuclear power reactor unless it finds that the state of onsite and offsite emergency preparedness provides reasonable assurance that adequate protective measures can and will be taken in a radiological emergency. The NRC bases its finding on a review of the DHS/FEMA findings and determinations as to whether State and local emergency plans are adequate and can be carried out, and on its own assessment of whether the onsite emergency plans are adequate and can be implemented (10 CFR 50.47(a)).

The principal guidance for preparing and evaluating radiological emergency plans for licensee and State and local government emergency planners is NUREG-0654/FEMA-REP-1, Revision 1, a joint NRC and FEMA document, issued November 1980. NUREG-0654 gives evaluation criteria for meeting the emergency planning standards in the NRC and FEMA regulations (10 CFR 50.47(b) and 44 CFR Part 350, "Review and Approval of State and Local Radiological Emergency Plans and Preparedness," respectively). These criteria provide a basis for licensees and State and local governments to develop acceptable emergency plans.

The NRC and DHS/FEMA coordinate their efforts in evaluating periodic emergency response exercises, which 10 CFR Part 50, Appendix E. IV. F.2, requires to be conducted every 2 years at all operating nuclear power plant sites. These full-participation exercises are integrated efforts by the licensee and State and local radiological emergency response organizations that have a role under the plan. The NRC evaluates the licensee's performance, and DHS/FEMA evaluates the response by State and local agencies. In some cases, various Federal response agencies also participate in these exercises. Any weaknesses or deficiencies identified by the NRC or DHS/FEMA as a result of the exercise must be corrected through appropriate remedial actions. Besides the biennial exercise of the plume exposure pathway plans, States must participate in an ingestion pathway exercise every 6 years with a nuclear power plant located within the States. There is no requirement to involve members of the public in any of the emergency preparedness exercises.

16.3 Emergency Classification System and Emergency Action Levels

The NRC regulations establish four classes of emergencies in order of increasing severity. Specifically, these are (1) unusual event, (2) alert, (3) site area emergency, and (4) general emergency. The specific class of emergency is declared on the basis of plant conditions that trigger the emergency action levels. Typically, licensees have established specific procedures for carrying out emergency plans for each class of emergency. The event classification initiates all appropriate actions for that class, including notification of offsite authorities, activation of onsite and offsite emergency response organizations, and, where appropriate, protective action recommendations for the public. These same emergency classes are also found in the State and local plans that support each nuclear power plant.

NUREG-0654 gives examples of initiating conditions for each of the four emergency classes. These conditions form the basis for each licensee to establish specific indicators, called emergency action levels. These levels provide a clear basis for rapidly identifying a possible problem and notifying the onsite emergency response organization and the offsite authorities that an emergency exists. Under NRC regulations, the licensee and State and local governmental authorities must discuss and agree upon the levels, and the NRC must approve them. In Regulatory Guide 1.101, "Emergency Planning and Preparedness for Nuclear Power Reactors," Revision 4, issued July 2003, the NRC endorsed the guidance in NUMARC/NESP-007, "Emergency Planning and Preparedness of Nuclear Power Plants," Revision 2, issued January 1992, and NEI 99-01, "Methodology for Development of Emergency Action Levels," Revision 4, issued January 2003, as acceptable alternatives for developing emergency action levels.

16.4 Recommendations for Protective Action in Severe Accidents

The technical basis and guidance for determining protective actions in the United States for severe (core damage) reactor accidents are given in NUREG-0654, Supplement 3, "Criteria for Protective Action Recommendations for Severe Accidents," Revision 1, issued July 1996, and EPA 400-R-92-001, "Manual of Protective Action Guides and Protective Actions for Nuclear Incidents," issued May 1992. These documents reflect the conclusions that have been developed from severe accident studies, such as NUREG-1150, "Severe Accident Risks: An Assessment for Five U.S. Nuclear Power Plants," issued December 1990.

Guidance for response procedures and training manuals for NRC staff appears in NUREG/BR-0150, "Response Technical Manual 96." The NRC's guidance on evacuation and sheltering in the event of a nuclear power plant accident is consistent with guidance in IAEA TECDOC-953, "Method for the Development of Emergency Response Preparedness for Nuclear or Radiological Accidents," and TECDOC-955, "Generic Assessment Procedures for Determining Protective Actions During a Reactor Accident," both issued in 1997. Additional generic communications have been issued regarding protective action recommendations.

The NRC considers evacuation and sheltering to be the two primary protective actions and prefers prompt evacuation for the population near a plant in a severe reactor accident. However, the NRC is currently evaluating this position, as under some circumstances, it may be better to shelter in place.

In addition, a supplemental protective action for the general population involves using the thyroid-blocking agent potassium iodide. The NRC amended its regulations for emergency planning (10 CFR 50.47(b)(o) in 2001. This amendment, "Consideration of Potassium Iodide in Emergency Plans," requires that each State consider giving potassium iodide to the general public as a protective measure, supplementing evacuation and sheltering. The NRC found that potassium iodide is a reasonable, prudent, and inexpensive supplement to evacuation and sheltering for specific local conditions. The NRC has funded an initial supply, as well as replenishment of expired potassium iodide tablets, for States that choose to give potassium iodide to the general public as part of their emergency plans. To date, 21 States have requested and received potassium iodide tablets. The NRC distributes 65-mg and 30-mg tablets. In January 2002, the NRC, in cooperation with the cognizant agencies, updated the Federal policy statement on potassium iodide prophylaxis to reflect the changes in NRC regulations. In September 2006, the Commission approved replenishment plans for initial State supplies.

16.5   Inspection Practices—Reactor Oversight Process for Emergency Preparedness

The NRC's Reactor Oversight Process, discussed in Article 6, addresses emergency preparedness. Specifically, the process allows the licensee latitude in managing emergency preparedness programs, including corrective actions, as long as the performance indicators and inspection findings are within an acceptable performance band. As explained in Article 6, the NRC handles inspection findings through its Significance Determination Process.

Emergency preparedness is the final barrier between reactor operations and protection of public health and safety. As such, emergency preparedness is a major component of the Reactor Oversight Process and is one of the seven recognized cornerstones of safety in the process. The objective established for this cornerstone is, "Ensure that the licensee is capable of implementing adequate measures to protect the public health and safety during a radiological emergency." Oversight of this cornerstone is achieved through three performance indicators and a supporting risk-informed inspection program. The performance indicators are drill and exercise performance, emergency response organization drill participation, and alert and notification system reliability. The performance indicator for drill and exercise performance monitors timely and accurate licensee performance in drills, exercises, and actual events when presented with opportunities to classify emergencies, notify offsite authorities, and recommend protective actions. The indicator for emergency response organization drill participation measures the percentage of key members of the licensee's emergency response organization who have participated in proficiency-enhancing drills, exercises, training opportunities, or an actual event over a certain time. The alert and notification system reliability indicator monitors the reliability of the offsite alert and notification system, which is a critical link for alerting and notifying the public of the need to take protective actions.

Under the Reactor Oversight Process, this cornerstone includes the following inspectable areas:

- Correction of Emergency Preparedness Weaknesses: Inspectors evaluate the licensees' programs for problem identification and resolution as they relate to emergency preparedness.

- Drill Evaluation: Inspectors evaluate drills and simulator-based training evolutions in which shift operating crews and licensee emergency response organization members participate.

- Exercise Evaluation: Inspectors independently observe the licensee's performance in classifying, notifying, and developing recommendations for protective actions, and other activities during the exercise. The inspectors also ensure that the licensee's critique is consistent with their observations.

- Alert and Notification System Evaluation: Inspectors verify the compliance of the testing program with program procedures.

- Emergency Action Level Changes: Inspectors review all of the licensee's changes to emergency action levels to determine if any of the changes have decreased the effectiveness of the emergency plan.

- Emergency Response Organization Staffing and Augmentation System: Inspectors review the augmentation system to determine whether, as designed, it will support augmentation of the emergency response organization in accordance with the goals for activating the emergency response facility.

- Reactor Safety—Emergency Preparedness: Inspectors verify that the data reported for the performance indicator values are valid.

- Emergency Action Level and Emergency Plan Changes: Inspectors sample changes to the emergency plan to ensure that the effectiveness of the emergency plan has not decreased.

- Force-on-Force Exercise Evaluation: Inspectors evaluate force-on-force exercises with respect to integration of security, plant operations, and emergency response. [Force-on-Force exercises assess a nuclear plants' physical protection to defend against the DBT. A full exercise, spanning several days, includes both table-top drills and simulated combat between a mock commando-type adversary force and the nuclear plant security force. The exercises are an essential part of NRC's oversight of plant owners' security programs and their compliance with NRC security requirements.]

Although DHS/FEMA has no direct regulatory authority over State and local governments, and the evaluators of FEMA exercises are not considered inspectors, the exercise findings of

115

DHS/FEMA carry substantial weight in the NRC's regulatory process. DHS/FEMA notifies the State government and the NRC of significant deficiencies in offsite performance shortly after the exercise, and DHS/FEMA issues a formal exercise report about 90 days after the exercise. This report describes the DHS/FEMA exercise findings, and the findings are expected to be closed either before or during the next exercise. Because of the potential effect of deficiencies on offsite emergency preparedness, they are expected to be corrected within 120 days of the exercise. Failure of offsite organizations to correct deficiencies in a timely manner could lead DHS/FEMA to withdraw its finding of "reasonable assurance."

16.6 Responding to an Emergency

Fundamental changes have occurred in the response to national emergencies as a result of the terrorist events of September 11, 2001, and Hurricane Katrina in August 2005. This section explains the roles of the Federal Government, licensees, State and local governments, and the NRC. It also explains the security aspects supporting the response.

16.6.1 Federal Response

The Federal response structure has been revamped with the creation of DHS and the implementation of Homeland Security Presidential Directive 5. This directive establishes the Secretary of Homeland Security as the primary Federal official for managing domestic incidents. Under the Homeland Security Act of 2002, DHS is responsible for coordinating Federal operations within the United States to prepare for, respond to, and recover from terrorist attacks, major disasters, and other emergencies.

Specifically, DHS will assure overall Federal incident management coordination responsibilities when any one of the following four conditions applies:

(1)   A Federal department or agency acting under its own authority has requested DHS assistance.

(2)   The resources of State and local authorities are overwhelmed, and the appropriate State and local authorities have requested Federal assistance.

(3)   More than one Federal department or agency has become substantially involved in responding to the incident.

(4)   The Secretary has been directed by the President to assume incident management responsibilities.

The framework that outlines the responsibilities of the Secretary of Homeland Security, DHS, and other Federal, State, and local entities is the National Response Plan, soon to be the National Response Framework, and its associated annexes. The framework provides guidance on Federal coordinating structures and processes to prepare for, respond to, and recover from domestic incidents such as terrorist attacks, major disasters, and other emergencies.

The Federal response to a potential nuclear/radiological incident is designed to support the efforts of the facility operator and offsite officials. For such emergencies, Federal response activities are conducted in accordance with the Nuclear/Radiological Incident Annex. The

Nuclear/Radiological Incident Annex describes the roles of DHS, coordinating agencies (i.e., the NRC in this type of emergency), and other supporting Federal agencies. During this type of incident, DHS is responsible for the overall domestic incident management, while the coordinating agency will coordinate the Federal on-scene actions and assist State and local governments in determining measures to protect life, property, and the environment. The coordinating agency may respond as part of the Federal response as requested by DHS under the framework, or in accordance with its own authorities. During less severe incidents, the coordinating agency will oversee the onsite response, monitor and support owner or operator activities (when there is an owner or operator), provide technical support to the owner or operator if requested, serve as the principal Federal source of information about onsite conditions, and, if requested, advise the State and local government agencies on implementing protective actions. The coordinating agency will also provide a hazard assessment of onsite conditions that might have significant offsite impact and ensure that onsite measures are taken to mitigate offsite consequences.

16.6.2 Licensee, State, and Local Response

The NRC recognizes the nuclear power plant operator (licensee) and the State or local government as the two primary decisionmakers in a radiological emergency at a licensed power reactor. The licensee is primarily responsible for mitigating the consequences of an incident on site and recommending timely and proper protective actions to State and local authorities. The State or local governments are ultimately responsible for implementing proper protective actions for public health and safety.

16.6.3 The NRC's Response

In fulfilling its legislative mandate for protecting the public health and safety, the NRC has developed a plan and procedures that detail its response to incidents involving licensed material and activities (NUREG-0728, "NRC Incident Response Plan," Revision 4, November 23, 2003). In accordance with that plan, the NRC will initially assess any reported event and decide whether or how it will respond as an agency. The NRC will generally dispatch a team to the site for all serious incidents to meet its statutory and regulatory obligations as the coordinating agency. The team may assist the State in interpreting and analyzing technical information while updating other responding Federal agencies on event conditions and coordinating any multiagency Federal response.

Once the NRC has decided to respond as an agency, the agency's Operations Center in the Washington, DC, area and the associated regional Incident Response Center are activated. The NRC Headquarters Operations Center will then (1) maintain continuous communications with the facility, (2) assess the incident, (3) advise the facility operator and offsite officials, (4) coordinate the Federal radiological response with other Federal agencies, and (5) respond to inquiries from the national media. The staff at the NRC Headquarters Operations Center includes emergency preparedness and response experts and personnel experienced with liaison activities. Early in an incident, the Regional Administrator provides operational authority from the affected regional office, and, if necessary, from the regional Incident Response Center because regional office personnel are usually most familiar with details of the affected facility. When a major NRC onsite presence is required, the NRC will dispatch a team from the affected regional office. The NRC Headquarters Operations Center will direct NRC response for about 4–8 hours until the lead is transferred to the NRC site team, if applicable.

As soon as the NRC site team arrives at the facility and is ready to assume the agency's leadership role, it is given certain authorities and responsibilities which may include the authority to direct the agency's response. The NRC site team then sends representatives to response centers that are used by the facility and offsite officials to coordinate the response. The NRC site team has access to extensive radiological monitoring capabilities through DOE, including field teams and aerial monitoring. The NRC site team also sends representatives to the joint information center established by the facility or local government to interact with the media.

The NRC regularly participates in exercises of its response program to ensure readiness to respond, participating in nuclear power plant, fuel cycle facility, and Federal interagency exercises each year. The NRC participates in the planning and conduct of the Top Officials (TOPOFF) exercises. The NRC's participation in such exercises gives the agency a valuable perspective on multievent response. This perspective improves interagency cooperation and imparts a better understanding of response roles during emergencies.

16.6.4 Security Aspects Supporting Response

Before September 11, 2001, the security measures at nuclear facilities provided reasonable assurance that the health and safety of the public would be protected in the event of an attack encompassed by the DBT of radiological theft and sabotage, which is described in 10 CFR 73.1, "Purpose and Scope." Since September 11, 2001, the nuclear industry has significantly enhanced its defensive capability as a result of the voluntary actions taken by licensees in response to the advisories issued by the NRC, and as required by the orders issued February 25, 2002, and January 7, 2003, and followed by the three orders issued April 29, 2003. The enhancements include security measures against threats from an insider, waterborne attack, vehicle bomb attack, and land-based assault. In addition, one of the orders issued April 29, 2003, identified a revised DBT against which licensees must be prepared to defend. The NRC is codifying through rulemaking (Article 6) many of the security requirements that were newly imposed on licensees by order following September 11, 2001. The NRC will consider additional measures in the future as necessary.

The NRC receives a substantial and steady flow of information from the national intelligence community, law enforcement, and licensees and continually evaluates this information to assess threats to regulated facilities or activities. The NRC works with a variety of other Federal agencies, particularly DHS and the Homeland Security Council, to ensure that security around nuclear power plants is well coordinated and that responders are prepared for a significant event. If an event were to occur, the NRC would coordinate the resources of more than 18 Federal agencies, to mitigate the radiological consequences of a serious accident or successful attack.

16.7 International Arrangements

The NRC has agreements with its neighbors, principally Canada and Mexico, and commitments to IAEA.

Under its signed agreements with Canada and Mexico, the NRC will promptly notify and exchange information in the event of an emergency that has the potential for trans-boundary effects. The agreement with Canada is the "Agreement Between the Government of the United States of America and the Government of Canada on Cooperation in Comprehensive Civil

Emergency Planning and Management." The procedure specified in "Administrative Arrangement Between the U.S. Nuclear Regulatory Commission and the Atomic Energy Control Board of Canada for Cooperation and the Exchange of Information in Nuclear Regulatory Matters" implements the agreement. (Both documents are dated June 21, 1989.)

The agreement with Mexico is the "Agreement for the Exchange of Information and Cooperation in Nuclear Safety Matters," which is implemented by the "Implementing Procedure for the Exchange of Technical Information and Cooperation in Nuclear Safety Matters Between the Nuclear Regulatory Commission of the United States of America and the Comision Nacional de Seguridad Nuclear y Salvaguardias of Mexico." (Both documents are dated October 6, 1989.)

To meet the U.S. commitment under the IAEA "Convention on Early Notification of a Nuclear Accident," the NRC will promptly notify IAEA if a serious accident occurs at a commercial nuclear power plant. Afterward, the NRC will work with the Department of State to update IAEA.

Since 2001, the United States has fully participated in the INES by evaluating operating reactor events and reporting to IAEA any events resulting in a categorization of INES Level 2 or higher.

ARTICLE 17. SITING

Each Contracting Party shall take the appropriate steps to ensure that appropriate procedures are established and implemented for

(i)   **evaluating all relevant site-related factors that are likely to affect the safety of a nuclear installation for its projected lifetime**

(ii)   **evaluating the likely safety impact of a proposed nuclear installation on individuals, society, and the environment**

(iii)   **re-evaluating, as necessary, all relevant factors referred to in subparagraphs (i) and (ii) so as to ensure the continued safety acceptability of the nuclear installation**

(iv)   **consulting Contracting Parties in the vicinity of a proposed nuclear installation, insofar as they are likely to be affected by that installation and, upon request, providing the necessary information to such Contracting Parties, in order to enable them to evaluate and make their own assessment of the likely safety impact on their own territory of the nuclear installation**

This section explains the NRC's responsibilities for siting, which include site safety, environmental protection, and emergency preparedness. First, this section discusses the regulations applying to site safety and their implementation. It emphasizes regulations applying to seismic, geological, and radiological assessments. Next, it explains environmental protection. Article 16 discusses emergency preparedness and international arrangements, which would apply to Contracting Parties in obligation (iv), above.

New information reported since the previous U.S. National Report includes early site permit review activities, new developments in seismic hazard analyses, the use of an alternative source term, and updated guidance.

17.1 Background

The NRC's siting responsibilities stem from the Atomic Energy Act, the Energy Reorganization Act (as discussed earlier), and the National Environmental Policy Act. These statutes confer broad regulatory powers on the Commission and specifically authorize the NRC to promulgate regulations that it deems necessary to fulfill its responsibilities under the acts.

The NRC's siting regulations are integral to protecting public health and safety and the environment. Siting away from densely populated centers has been, and will continue to be, an essential component of the NRC's defense-in-depth safety philosophy (see Article 18), which also includes multiple-barrier containment and redundant safety systems. The primary factors that determine public health and safety are the reactor design and construction and operation of the facility. However, siting factors and criteria are important in ensuring that radiological doses from normal operation and postulated accidents will be acceptably low, natural phenomena and man-made hazards will be properly accounted for in the design of the plant, and the human environment will be protected during the construction and operation of the plant.

For the first time since the 1970s, the nuclear power industry in the United States is seeking approval for sites that could host new nuclear power plants. To ensure that the agency can effectively carry out its responsibilities associated with, among others, an early site permit application, the NRC consolidated regulatory functions to (1) manage near-term future licensing activities, (2) work with stakeholders regarding new reactor licensing activities, and (3) assess the NRC's readiness to perform new reactor licensing reviews.

In 2003, applicants submitted three early site permit applications to the NRC for sites in Virginia, Illinois, and Mississippi. In 2006, an applicant submitted an early site permit application for a site in Georgia. The sites are in proximity to existing nuclear power plants, which enables the applicants to use existing physical and administrative infrastructures and existing programs and siting information and to reduce the impact on the environment compared to the impact a plant would have on an undeveloped location.

In anticipation of these applications and to ensure that future license applicants and the public understand the NRC's review process of programs and siting information, the NRC documented its review process and criteria in RS-002, "Processing Applications for Early Site Permits," issued December 2003.

The NRC expects to receive an unprecedented number of applications that require siting evaluations principally under the combined license application provisions of 10 CFR Part 52. While many of these applications will be for locations close to existing facilities, some will be at locations where applicants requested construction permits under 10 CFR Part 50 but plants were not completed, and yet others will be at previously undeveloped ("green field") sites.

17.2 Safety Elements of Siting

This section explains the safety elements of siting. After providing a short background, it explains seismic and geological assessments. It then discusses radiological assessments performed for initial licensing, as a result of facility changes, and according to regulatory developments that have occurred since the licensing of all U.S. operating plants.

17.2.1 Background

The NRC's site safety regulations consider societal and demographic factors, manmade hazards (such as airports and dams), and physical characteristics of the site (such as seismic and meteorological factors) that could affect the design of the plant. The requirements are specified in 10 CFR Part 100, "Reactor Site Criteria"; Appendix A, "Seismic and Geologic Siting Criteria for Nuclear Power Plants," to 10 CFR Part 100; Subpart B, "Evaluation Factors for Stationary Power Reactor Site Applications on or after January 10, 1997," of 10 CFR Part 100; and 10 CFR 100.23, "Geologic and Seismic Siting Criteria." The requirements in 10 CFR 100.23 apply to applicants for an early site permit, a combined license, a construction permit, or an operating license on or after January 10, 1997. Regulatory Guides 1.165, "Identification and Characterization of Seismic Sources and Determination of Safe Shutdown Earthquake Ground Motion," issued March 1997, and 1.208, "A Performance-Based Approach to Define the Site-Specific Earthquake Ground Motion," issued March 2007, describe methods acceptable to NRC staff for implementing those requirements, and NUREG-0800, Section 2.5.2, Revision 3, guides the staff in its reviews.

The applicant's safety analysis report must describe characteristics in and around the site and contain accident analyses that are relevant to evaluating the suitability of a site. A number of regulatory guides provide guidance regarding issues of site safety that applicants need to address. NUREG-0800 guides the staff in reviewing the site safety content of these reports. RS-002 identifies parts of NUREG-0800 that apply to the review of early site permits.

Once licensed to operate, the licensee is expected to monitor the environs around the nuclear power plant and report changes in the environs in its safety analysis report that may affect the continued safe operation of the facility.

17.2.2 Assessments of Seismic and Geological Aspects of Siting

The siting regulations stated in Section 17.2.1 above detail the assessments applying to seismic and geologic aspects of siting. More recent developments in assessments include the performance-based approach for determining the site-specific ground motion response spectrum and the safe-shutdown earthquake. The performance-based approach combines the site seismic hazard curves and seismic fragility curves for nuclear structures to meet a specified performance target. RG 1.208, which was developed as an alternative to Regulatory Guide 1.165, describes this new approach in detail.

Regulatory Guide 1.208 also incorporates recent developments in the area of seismic hazard assessment. These recent developments include the use of cumulative absolute velocity filtering in place of a lower-bound magnitude cutoff, as well as guidance on the development of earthquake time histories, site response analysis, and the location of the ground motion response spectrum within the soil profile.

In 2003, the three early site permit applicants used the EPRI Central and Eastern United States (CEUS) seismic source models as a starting point for their site applications. Applicants updated the EPRI source models to reflect advances in CEUS seismic and geologic source modeling. In addition, EPRI updated its ground motion models for generic use in new plant probabilistic seismic hazard analyses for sites located in the CEUS in 2003.

Advanced reactor designs are reviewed and certified under 10 CFR Part 52, and they use high seismic design input that is independent of any site, but are capable of being sited in majority of currently existing sites. All new and advanced reactor designs are required to demonstrate that they have a plant level seismic margin of 1.67 times the design basis safe shutdown earthquake with high confidence (95%) in low (5%) probability of failure.

In summary, new seismic demand for design of new reactors ensures that the frequency at which nuclear structures, systems and components will reach the threshold of elastic limits under seismic loads combined with dead, live and postulated accident loads is 10^{-5} per reactor year. Hence the margin of a plant to failure under a design basis seismic events is greater than 1.67.

17.2.3 Assessments of Radiological Consequences

The Reactor Site Criteria Rule, 10 CFR Part 100, is the regulation under which all U.S. operating plants were licensed. It contains provisions for assessing whether radiological doses from postulated accidents will be acceptably low. The NRC has issued the following regulatory guidance for licensees to implement the requirements regarding the radiological criteria of

10 CFR Part 100:

- Regulatory Guide 1.3, "Assumptions Used for Evaluating the Potential Radiological Consequences of a Loss-of-Coolant Accident for Boiling-Water Reactors," Revision 2, June 1974

- Regulatory Guide 1.4, "Assumptions Used for Evaluating the Potential Radiological Consequences of a Loss-of-Coolant Accident for Pressurized-Water Reactors," Revision 2, June 1974

- Regulatory Guide 1.145, "Atmospheric Dispersion Models for Potential Accident Consequence Assessments at Nuclear Power Plants," Revision 1, November 1982

Although applicants perform dose analyses primarily to support reactor siting, licensees are required to evaluate the potential increase in the consequences of accidents that might result from modifying facility SSCs. Commitments (including the radiological acceptance criteria) made by the applicant during siting and documented in its final safety analysis report remain binding until modified. Consequently, a licensee must evaluate the potential consequences of design changes against these radiological criteria to demonstrate that the design changes result in a design that still conforms to the regulations and commitments. If the consequences increase more than minimally, as outlined in 10 CFR 50.59 (or require a change to the technical specifications), as discussed in Article 14, the licensee must obtain NRC approval before implementing the proposed modification.

There have been regulatory developments since the licensing of all U.S. plants now operating. These include a revision to 10 CFR Part 100 in 1996; NUREG-1465, "Accident Source Terms for Light-Water Nuclear Power Plants" issued February 1995; Regulatory Guide 1.183, "Alternative Radiological Source Terms for Evaluating Design Basis Accidents at Nuclear Power Reactors," issued July 2000, which guided the use of NUREG-1465; and 10 CFR 50.67, "Accident Source Term," which allowed licensees to use alternative source terms. (The previous U.S. National Report discussed these developments.)

The NRC has applied the 1996 revision to 10 CFR Part 100, along with the alternative source term, in its design certification review for a passive advanced light-water reactor, the AP600. More recently, the agency has applied the practice to the AP1000 with similar results and is expected to apply it for all contemplated light-water reactors, including the economic and simplified boiling water reactor (ESBWR) design certification review. For other-than-light-water reactor designs, applicants will have to describe their rationale for an appropriate accident source term characterization which will be subject to NRC independent review.

The industry continues to explore the use of the alternative source term in implementing cost-beneficial licensing actions at operating reactors. Some of these applications resulted in improved safety equipment reliability and in reduced occupational exposures. Since the issuing of 10 CFR 50.67 more than half of the operating reactor licensees requested either full implementation of the alternative source term or selective implementation for certain regulatory applications. Operating plant licensees have also used the alternative source term to analyze the adequacy of certain engineered safety features in meeting the operability requirements in their operating reactor technical specifications.

17.3 Environmental Protection Elements of Siting

This section explains the environmental protection elements of siting. It covers the governing documents and site approval process. Since the last operating plants in the United States received licenses, issues have arisen that must be considered in siting reviews. This section explains the effect of these issues on siting reviews.

17.3.1 Governing Documents and Process

The environmental protection elements of siting consist of the plant's demands on the environment (e.g., water use and effects of construction and operation). These elements are addressed in 10 CFR Part 51, which implements the National Environmental Policy Act, consistent with the NRC's statutory authority, and reflects the agency's policy to voluntarily apply the regulations of the President's Council on Environmental Quality, subject to certain conditions. Integrating environmental reviews into its routine decisionmaking, the NRC considers environmental protection issues and alternatives before taking any action that may significantly affect the human environment.

The site approval process leading to the construction or operation of a nuclear power plant requires the NRC to prepare an environmental impact statement. The updated and revised environmental standard review plans (NUREG-1555) guide the staff's environmental reviews for a range of applications, including green field (i.e., undeveloped sites) reviews for construction permits and operating licenses in 10 CFR Part 50, for early site permits in 10 CFR Part 52, Subpart A, "Early Site Permits," and for combined licenses in 10 CFR Part 52, Subpart C, "Combined Licenses," when the application does not reference an early site permit. Article 19, in Regulatory Guide 1.206, "Combined Operating Licenses for Nuclear Power Plants," and RS-002, dealing with early site permits, discuss these governing documents and processes. Environmental standard review plans are also appropriate for environmental reviews of applications for combined licenses in 10 CFR Part 52, Subpart C, when the applications reference an early site permit. Reviews of early site permit applications are limited in the sense that (1) the reviews focus on the environmental effects of reactor construction and operation that have characteristics that fall within the postulated site parameters and (2) the reviews need not assess benefits (e.g., the need for power). The environmental information in applications for combined licenses that reference an early site permit is limited to consideration of (1) information to demonstrate that the design of the facility falls within the parameters specified in the early site permit, (2) new and significant information on issues previously considered in the early site permit proceeding, and (3) any significant environmental issue not considered in any previous proceeding on the site or design.

The environmental standard review plans in Supplement 1 to NUREG-1555 guide the staff's environmental review for license renewal applications under 10 CFR Part 54, which is discussed in Article 14.

Several other NRC actions on siting and site suitability require environmental reviews, including issuance of limited work authorizations (10 CFR 50.10(e)(1) to (e)(3), 10 CFR 52.25, "Extent of Activities Permitted," and 10 CFR 52.91, "Authorization to Conduct Site Activities"), early partial decisions (10 CFR 2.600, "Scope of Subpart," in Subpart F, "Additional Procedures Applicable to Early Partial Decisions on Site Suitability Issues in Connection with an Application for a Permit to Construct Certain Utilization Facilities," of 10 CFR Part 2), and preapplication early reviews of site suitability issues (Appendix Q, "Preapplication Early Review of Site Suitability Issues," to 10 CFR Part 50).

17.3.2 Other Considerations for Siting Reviews

Since the NRC last issued construction permits under 10 CFR Part 50 in the 1970s, and coincident with the publication of the initial environmental standard review plan, many changes to the regulatory environment have affected the NRC and applicants seeking site approvals. These include new environmental laws and regulations, changes in policies and procedures resulting from decisions of courts and administrative hearing boards, and changes in the types of authorizations, permits, and licenses issued by the NRC. The following paragraphs highlight some of these changes and their effects on the environmental standard review plans.

In the late 1980s, the NRC issued regulations that provided an alternative licensing framework to 10 CFR Part 50, which provided for a construction permit followed by an operating license. The new framework provided in 10 CFR Part 52 introduced the concept of approving designs independent of sites, and approving sites independent of designs, and then efficiently linked the approvals to result in the approval to construct and operate the facility. As discussed earlier, the NRC has received four early site permit applications under 10 CFR Part 52 and is actively conducting siting reviews.

Toward that end, the NRC issued RS-002, which embodies the environmental guidance in NUREG-1555, the environmental standard review plan, and the outcome of interactions with stakeholders. In addition, the NRC is revising 10 CFR Part 52 to reflect experience gained in its use and to provide guidance on the preparation of combined license applications, including guidance on environmental issues, in RG 1.206.

As described in previous U.S. National Reports, other relevant regulatory developments include:

- Presidential Executive Order 12898, "Federal Actions to Address Environmental Justice in Minority and Low-Income Populations," issued February 1994, which instructed Federal agencies to make "environmental justice" part of each agency's mission by addressing disproportionately high and adverse human health or environmental effects of Federal programs, policies, and activities on minority and low-income populations

- the Yellow Creek Decision, which determined that the authority of the NRC is limited in matters that are expressly assigned to EPA

- changes in the economic regulation of utilities that have expanded the options to be addressed in considering the need for power in environmental impact statements

- design alternatives to mitigate the consequences of severe accidents

ARTICLE 18. DESIGN AND CONSTRUCTION

Each Contracting Party shall take the appropriate steps to ensure that:

(i)   the design and construction of a nuclear installation provides for several reliable levels and methods of protection (defense in depth) against the release of radioactive materials, with a view to preventing the occurrence of accidents and to mitigating their radiological consequences should they occur

(ii)   the technologies incorporated in the design and construction of a nuclear installation are proven by experience or qualified by testing or analysis

(iii)   the design of a nuclear installation allows for reliable, stable, and easily manageable operation, with specific consideration of human factors and the man-machine interface

This section explains the defense-in-depth philosophy and how it is embodied in the general design criteria of U.S. regulations. It explains how applicants meet the defense-in-depth philosophy and how the NRC reviews applications and conducts inspections before issuing licenses to ensure that this philosophy is implemented in practice. Next, this section discusses measures for ensuring that the applications of technologies are proven by experience or qualified by testing or analysis. Section 14.2 of this report also addresses this obligation. Finally, this section discusses requirements regarding reliable, stable, and easily manageable operation, specifically considering human factors and the man-machine interface. Article 12 also addresses this obligation.

The changes reported since the previous U.S. National Report are an updating of design certifications that are either completed or under review, governing documents, and experience.

18.1 Defense-in-Depth Philosophy

This section explains the defense-in-depth philosophy followed in regulatory practice and the governing documents and regulatory process relevant to designing and constructing a nuclear power plant. It also discusses relevant experience and examples.

18.1.1 Governing Documents and Process

The defense-in-depth philosophy, as applied in regulatory practice, requires that nuclear plants contain a series of independent, redundant, and diverse safety systems. The physical barriers for defense in depth in a light-water reactor are the fuel matrix, the fuel rod cladding, the primary coolant pressure boundary, and the containment. The levels of protection in defense in depth are (1) a conservative design, QA, and safety culture, (2) control of abnormal operation and detection of failures, (3) safety and protection systems, (4) accident management, including containment protection, and (5) emergency preparedness.

Appendix A to 10 CFR Part 50 embodies the defense-in-depth philosophy. General design criteria cover protection by multiple fission product barriers, protection and reactivity control systems, fluid systems, containment design, and fuel and radioactivity control. The NRC staff amplified its defense-in-depth philosophy in Regulatory Guide 1.174, which provides guidance on using a PRA in risk-informed decisions on plant-specific changes. The general design criteria establish the minimum requirements for the principal design criteria, which in turn establish the necessary design, fabrication, construction, testing, and performance requirements for SSCs that are important to safety.

To ensure that a plant is properly designed and built as designed, that proper materials are used in construction, that future design modifications are controlled, and that appropriate maintenance and operational practices are followed, a good QA program is needed. To meet this need, General Design Criterion 1 of Appendix A to 10 CFR Part 50 and its implementing regulatory requirements specified in Appendix B to 10 CFR Part 50 establish QA requirements for all activities affecting the safety-related functions of the SSCs.

Pursuant to the two-step licensing process set forth in 10 CFR Part 50, an applicant for a construction permit must present the principal design criteria for a proposed facility in its preliminary safety analysis report (see 10 CFR 50.34). For guidance in writing a safety analysis report, the applicant may use Regulatory Guide 1.70. The safety analysis report must also contain design information for the proposed reactor and comprehensive data on the proposed site. The report must also discuss various hypothetical accident situations and the safety features to prevent accidents or, if accidents occur, to mitigate their effects on both the public and the facility's employees. After obtaining a construction permit under 10 CFR Part 50, the applicant must submit a final safety analysis report to support an application for an operating license, unless it submitted the report with the original application. This report gives the details of the final design of the facility, plans for operation, and procedures for coping with emergencies. The preliminary and final safety analysis reports are the principal documents that the applicant provides for the staff to determine whether the proposed plant can be built and operated without undue risk to the health and safety of the public. The NRC expects that future applications to build nuclear power plants will use the combined license process under 10 CFR Part 52. Applications submitted under Part 52 must meet all of the Part 50 requirements. A significant difference in the Part 52 process is that the final safety analysis report must be submitted before authorization is granted to begin construction. Article 19 discusses the combined license review process.

The NRC staff reviews safety analysis reports according to NUREG-0800 to ensure that the applicant has satisfied the general design criteria and other applicable regulations. The staff reviews each application to determine whether the plant design meets the Commission's regulations (10 CFR Part 20, 10 CFR Part 50, 10 CFR Part 73, and 10 CFR Part 100). These reviews include, in part, the characteristics of the site. In addition, each application for a nuclear installation must have a comprehensive environmental report that provides a basis for evaluating the environmental impact of the proposed facility. Regulatory Guide 4.2, "Preparation of Environmental Reports for Nuclear Power Stations," Revision 2 issued July 1976 provides applicants with information on writing environmental reports. The NRC staff reviews the environmental reports according to NUREG-1555. In reviewing an applicant's submittal, the staff, supported by outside experts, conducts independent technical studies to review certain safety and environmental matters. The staff states its conclusions in an environmental impact statement and a safety evaluation report, which it may update before granting the license. Under the two-step licensing process in 10 CFR Part 50, the NRC does not issue an operating license until construction is complete and the Commission makes the findings set forth in 10 CFR 50.57, "Issuance of Operating License." For applications submitted under

10 CFR Part 52, the Commission must find that all acceptance criteria in the combined license are met prior to operation of the facility.

The NRC maintains surveillance over nuclear power plant construction to ensure compliance with the agency's regulations to protect public health and safety and the environment. The NRC's inspection program has been anticipating that future applicants for construction of a nuclear power plant will apply for a combined license under 10 CFR Part 52. The NRC has developed an inspection program for future nuclear plants licensed under 10 CFR Part 52.

The new inspection program revises the 10 CFR Part 50 construction inspection program. It incorporates inspections, tests, analyses, and acceptance criteria (ITAAC) from 10 CFR Part 52, as well as lessons learned from the inspection program used in the previous construction era (1970–1980), and considers modular construction at remote locations.

Before construction, the NRC inspection program focuses on the applicant's establishing a QA program to verify that applications submitted to the NRC meet specified requirements in 10 CFR Part 52 and are of a quality suitable for docketing. Inspection Manual Chapter 2501, "Early Site Permit," lists inspections for this phase.

Once the NRC receives an application, the inspection program focuses on supporting the NRC staff's preparation for the mandatory Atomic Safety and Licensing Board hearing and the final Commission decision on whether a combined license should be granted. Inspection Manual Chapter 2502, "Pre-Combined License Phase," lists inspections for this phase.

During construction, inspectors sample the spectrum of the applicant's activities related to performance of the ITAAC in the design-basis document to confirm that the applicant is adhering to quality and program requirements. NRC inspectors will verify successful ITAAC completion on a sampling basis and will review all ITAACs. The NRC will publish notices in the *Federal Register* of those ITAACs that have been completed. Inspection Manual Chapter 2503, "ITAAC," lists inspections for this phase.

As the applicant completes construction, the inspection program focuses on verifying the adequacy of the licensee's preoperational programs such as fire protection, security, training, radiation protection, startup testing, and programs that enable the transition of the organization from construction to power operations. Inspection Manual Chapter 2504, "Non-ITAAC Inspections," lists inspections for this phase.

18.1.2 Experience

18.1.2.1 Regulatory Framework for the Reactivation of Watts Bar Unit 2

The Watts Bar Nuclear Plant (WBN) is located in southeastern Tennessee and is owned by the TVA. The site has two Westinghouse designed PWRs. WBN Unit 1 received a full power operating license in early 1996, and was the last power reactor that was licensed in the United States. TVA stopped construction activities at WBN Unit 2 in mid 1980s. TVA is planning to resume WBN 2 construction and pursue operating license approval under 10 CFR Part 50. The construction permit for WBN 2 is currently active and expires in 2010.

TVA initiated a study of the feasibility of resuming construction of WBN Unit 2 with a planned start of the facility by 2013. By letter dated August 3, 2007, TVA notified the Director of NRR 120 days in advance of the reactivation of construction in accordance with the Commission Policy Statement on Deferred Plants.

The NRC will perform necessary regulatory review before issuance of operating license for WBN Unit 2. The regulatory framework for the potential reactivation WBN Unit 2 will include inspection and licensing activities.

18.1.2.2 Design Certifications

For more than 30 years, the Atomic Energy Commission and the NRC have reviewed applications submitted under the two-step licensing process in 10 CFR Part 50 and documented their reviews in safety evaluation reports and their supplements for 110 nuclear installations. Subsequently, the NRC has certified four standard plant designs under the design certification process in 10 CFR Part 52—General Electric's advanced BWR (1997), and Westinghouse's System 80+ (designed and licensed by Combustion Engineering), AP600, and AP1000 (1997, 2000, and 2006, respectively). General Electric's ESBWR design is currently under review for design certification.

18.2 Technologies Proven by Experience or Qualified by Testing or Analysis

The earlier discussions in Section 18.1.1 and Section 14.2 address the qualification of currently used technologies. The NRC ensures that new technologies are proven as required by 10 CFR 52.47(b). This rule requires demonstration of new technologies through analysis, appropriate test programs, experience, or a combination thereof. Most recently, Westinghouse used separate effects tests, integral systems tests, and analyses to demonstrate that its passive safety systems will perform as predicted in its safety analysis reports for the AP600 and AP1000 standard plant designs.

18.3 Design for Reliable, Stable, and Easily Manageable Operation

The NRC specifically considers human factors and the human-system interface in the design of nuclear installations. For safety analysis reports, the NRC reviews the human factors engineering design of the main control room and the control centers outside of the main control room. Article 12 also discusses human factors.

18.3.1 Governing Documents and Process

To support its reviews of the human factors engineering issues associated with the certification and licensing of new plant designs, the NRC uses Revision 1 of Chapter 18 of NUREG-0800, and Revision 2 of NUREG-0700, "Human-System Interface Design Review Guideline," issued May 2002. The NRC also uses Revision 2 of NUREG-0711 for evaluating the design of next-generation main control rooms. NUREG-0800, Section 14.3.9 provides additional guidance. Moreover, the NRC developed a new guidance document for use in reviewing combined license applications; that document, Regulatory Guide 1.206, "Combined License Applications for Nuclear Power Plants (LWR Edition)," includes sections that address the human factors engineering review of combined license applications.

18.3.2 Experience

The NRC's recently formed Office of New Reactors is actively reviewing new plant designs and preparing for the review of combined license applications. The NRC is currently conducting a design certification review of General Electric's ESBWR and is reviewing preapplication documents submitted by vendors who anticipate filing applications with the NRC in the future.

18.3.2.1 Digital Instrumentation and Controls

In recent years, nuclear facility and byproduct licensees have begun replacing their analog instrumentation and control (I&C) safety systems and equipment with digital systems and equipment. While digital technology has the capability to improve operational performance, the introduction of this technology into nuclear facilities and applications poses a variety of challenges for the NRC and the nuclear industry. In particular, these challenges include (1) the increased complexity of digital technology compared to analog technology; (2) rapid changes in digital technology that require the NRC to update its knowledge of the state-of-the-practice in digital system design, testing, and application; (3) new failure modes associated with digital technology; and (4) the need to update the acceptance criteria and review procedures used in consistently assessing the safety and security of digital systems. In response to these technical challenges, in January 2007, the NRC formed a digital I&C steering committee. The steering committee will provide management focus on the NRC regulatory activities in progress across several offices, interface with the industry on key issues, and facilitate consistent approaches to resolving technical and regulatory challenges. The members of the steering committee include management representatives from the various NRC offices that have regulatory responsibilities related to digital I&C.

Digital instrumentation and controls raises issues that were not relevant to analog systems. Examples of such issues include the following:

1. A common-cause failure attributable to software errors was not possible with analog systems. This potential weakness may require a consideration of diversity and defense-in-depth in the application of digital I&C systems.

2. Digital system network architectures also raise issues such as interchannel communication, communication between nonsafety and safety systems, and cyber security that must be reviewed closely to ensure that public safety is preserved.

3. Highly integrated control room designs with safety and nonsafety displays and controls will be the norm for new reactor designs. Human Factors design and Quality assurance during all phases of software development, control, and validation and verification are critical.

The Digital Instrumentation and Control steering committee has formed the following six task working groups that focus on key areas of concern:

- Cyber Security

- Diversity and Defense-in-Depth

- Risk-Informed Digital I&C

- Highly-Integrated Control Room - Communications

- Highly-Integrated Control Room - Human Factors

- Licensing Process Issues

Additionally, as directed by the Commission, the NRC staff is planning a public workshop to explore the feasibility of an integrated digital instrumentation and control and human-machine interface test facility. The staff is involving stakeholders in other government agencies, the national labs, industry, vendors and universities.

18.3.2.2 Cyber Security

After September 11, 2001, the NRC issued two security-related orders to require power reactor licensees to implement measures to enhance cyber security. These security measures required an immediate identification and assessment of computer-based systems deemed to be critical to the operation and security of the facility. Additionally, licensees were expected to implement any immediate and necessary corrective measures to protect against the cyber threats at the time the orders were issued.

Recognizing that licensees likely used various approaches in the architectural design and implementation of plant computing networks, the NRC embarked upon an effort to develop a cyber security self-assessment methodology that could be uniformly applied to U.S.-based nuclear facilities. Development of such a methodology would provide a means to ensure that the assessments performed by each facility would follow a consistent, repeatable approach, thereby providing comparable metrics to understand the relative cyber security posture of each facility. The assessment methodology was developed by a multidisciplinary team from Pacific Northwest National Laboratory with input from the NRC and nuclear power industry representatives and issued as NUREG/CR-6847 "Cyber security Self-Assessment Method for U.S. Nuclear Power Plants."

Using NUREG/CR-6847 as a foundation, the Nuclear Energy Institute (NEI) Cyber Security Task Force developed a comprehensive guidance document, NEI 04-04, "Cyber Security Programs for Power Reactors," that licensees can use to develop and manage an effective cyber security program. In December 2005, the NRC staff accepted NEI 04-04 as an acceptable method for establishing and maintaining a cyber security program at nuclear power plants.

In parallel with the development effort of NEI 04-04, the NRC revised existing regulatory guidance on use of computers in nuclear digital safety systems. In addition, the NRC has implemented a significant and continuing research program in cyber security for digital plant control systems. Finally, it is codifying the mandated cyber security enhancement requirements in the two security-related NRC orders by amending its regulations.

132

ARTICLE 19. OPERATION

Each Contracting Party shall take appropriate steps to ensure that:

(i)   **the initial authorization to operate a nuclear installation is based upon an appropriate safety analysis and a commissioning program demonstrating that the installation, as constructed, is consistent with design and safety requirements**

(ii)   **operational limits and conditions derived from the safety analysis, test, and operational experience are defined and revised as necessary for identifying safe boundaries for operation**

(iii)   **operation, maintenance, inspection, and testing of a nuclear installation are conducted in accordance with approved procedures**

(iv)   **procedures are established for responding to anticipated operational occurrences and to accidents**

(v)   **necessary engineering and technical support in all safety related fields is available throughout the lifetime of a nuclear installation**

(vi)   **incidents significant to safety are reported in a timely manner by the holder of the relevant license to the regulatory body**

(vii)   **programs to collect and analyze operating experience are established, the results obtained and the conclusions drawn are acted upon and that existing mechanisms are used to share important experience with international bodies and with other operating organizations and regulatory bodies**

(viii)   **the generation of radioactive waste resulting from the operation of a nuclear installation is kept to the minimum practicable for the process concerned, both in activity and in volume, and any necessary treatment and storage of spent fuel and waste directly related to the operation and on the same site as that of the nuclear installation take into consideration conditioning and disposal**

The NRC relies on regulations in Title 10, "Energy," of the *Code of Federal Regulations* and internally developed associated programs in granting the initial authorization to operate a nuclear installation and in monitoring its safe operation throughout its life. The material that follows describes the more significant regulations and programs corresponding to each obligation of Article 19.

This update discusses the revised Operating Experience Program.

19.1 Initial Authorization to Operate

All currently operating reactors in the United States received licenses under the two-step process in 10 CFR Part 50. This licensing process requires both a construction permit and an operating license. The additional licensing processes in 10 CFR Part 52 provide for site approvals and design approvals in advance of construction authorization. In addition, 10 CFR Part 52 includes a process that combines a construction permit and an operating license with conditions into one license (combined license). Both the two-step and the combined license processes require NRC approval to construct and operate a nuclear power plant.

The Advisory Committee on Reactor Safeguards, an independent statutory committee established to advise the NRC on reactor safety, reviews each application to construct or operate a nuclear power plant. The committee begins its review early in the licensing process by selecting proper stages at which to hold a series of meetings with the applicant and NRC staff. Upon completing its review, the committee reports to the Commission.

The public also has an opportunity to have its concerns addressed. The Atomic Energy Act requires that a public hearing be held before a construction permit, early site permit, or a combined license may be issued for a nuclear power plant. A three-member Atomic Safety and Licensing Board, which consists of one lawyer who acts as chairperson and two technically qualified persons, conducts the public hearing. Members of the public may submit statements to the licensing board, or they may petition for leave to intervene as full parties in the hearing.

To obtain NRC approval to construct or operate a nuclear power plant, an applicant must submit safety analysis reports. Article 18 describes the final safety analysis report and the NRC's review of the application for an operating license. A public hearing is neither mandatory nor automatic for an application for an operating license under 10 CFR Part 50. However, soon after the NRC accepts the application for review, it publishes a notice that it is considering issuing the license. This notice states that any person whose interest might be affected by the proceeding may petition the NRC for a hearing. If a public hearing is held, the same process described for the hearing for the construction permit applies.

A combined license, issued under Subpart C of 10 CFR Part 52, authorizes construction of a facility in a manner similar to a construction permit under 10 CFR Part 50. Just as for a construction permit, the NRC must hold a hearing before the decision on issuance of a combined license. However, the combined license will specify the inspections, tests, and analyses that the licensee must perform and the acceptance criteria that, if met, are necessary and sufficient to provide reasonable assurance that the facility has been constructed and will be operated in conformity with the license and the applicable regulations. After issuing a combined license, the NRC staff will verify that the licensee has performed the required inspections, tests, and analyses, and before operation of the facility, the Commission must find that the licensee has met the acceptance criteria. At periodic intervals during construction, the NRC staff will publish notices of the successful completion of inspections, tests, and analyses in the *Federal Register*. Then, not less than 180 days before the date scheduled for initial loading of fuel, the NRC will publish a notice of intended operation of the facility in the *Federal Register*. An opportunity for a second hearing exists, but petitions for this hearing will be considered only if the petitioner demonstrates that one or more of the acceptance criteria have not been (or will not be) met, and the specific operational consequences of nonconformance would be contrary to providing reasonable assurance of adequate protection of the public health and safety.

An early site permit, issued under Subpart A of 10 CFR Part 52, provides for resolution of site safety, environmental protection, and emergency preparedness issues, independent of a specific nuclear plant design review. The application for an early site permit must address the safety and environmental characteristics of the site, and evaluate potential physical impediments to the development of an acceptable emergency plan or security plan. Additional detail may be submitted on emergency preparedness issues up to a complete emergency plan. The staff documents its findings on site safety characteristics and emergency planning in a safety evaluation report and findings on environmental protection issues in an environmental impact statement. The early site permit may also allow nonsafety site preparation activities, subject to redress, before the issuance of a combined license. The NRC will issue a *Federal Register* notice for a mandatory public hearing, and the Advisory Committee on Reactor Safeguards will perform an independent safety review. A construction permit or combined license application may reference the early site permit.

Under Subpart B, "Standard Design Certifications," of 10 CFR Part 52, the NRC may certify and approve a standard plant design through a rulemaking, independent of a specific site. The issues resolved in a design certification have a more restrictive backfit requirement than issues resolved under other licenses. That is, the NRC cannot modify a certified design unless the modification is necessary to meet the applicable regulations in effect during design certification, or to ensure adequate protection of public health and safety. An application for a combined license under 10 CFR Part 52 can incorporate by reference a design certification, an early site permit, or both. The advantage of this approach is that the issues resolved by rulemaking for design certification and those resolved during the early site permit hearing process are precluded from reconsideration at the combined license stage.

19.2 <u>Definition and Revision of Operational Limits and Conditions</u>

The license for each nuclear facility must contain technical specifications that set operational limits and conditions derived from the safety analyses, tests, and operational experience. The regulations in 10 CFR 50.36 define the requirements that apply to the plant-specific technical specifications. At a minimum, the technical specifications must describe the specific characteristics of the facility and the conditions for its operation that are required to adequately protect the health and safety of the public. Each applicant must identify items that directly apply to maintaining the integrity of the physical barriers that are designed to contain radioactive material. Specifically, 10 CFR 50.36 requires that the technical specifications must be derived from the analyses and evaluation in the safety analysis report. Licensees cannot change the technical specifications without prior NRC approval.

In 1992, the NRC issued improved vendor-specific (e.g., Babcock and Wilcox, Westinghouse, Combustion Engineering, and General Electric) standard technical specifications in NUREGs 1430-1434, and periodically revises them on the basis of experience. The NRC issued Revision 3 to these NUREGs in June 2004.

The NRC encourages licensees to use the improved standard technical specifications as the basis for plant-specific technical specifications. The agency also considers requests to adopt parts of the improved standard technical specifications, even if the licensee does not adopt all of the improvements. These parts, which will include all related requirements, will normally be developed as line-item improvements. To date, over half of the operating commercial nuclear plants have converted their technical specifications to the improved standard technical specifications.

Consistent with the Commission's policy statements on technical specifications and the use of PRA, the NRC and the nuclear industry are developing risk-informed improvements to technical specifications. These improvements, or initiatives, are intended to maintain or improve safety while reducing unnecessary burden and to bring technical specifications into congruence with the agency's other risk-informed regulatory requirements (in particular, the risk management requirements of the Maintenance Rule in 10 CFR 50.65(a)(4)).

19.3 Approved Procedures

In the United States, operations, maintenance, inspection, and testing of a nuclear installation are conducted in accordance with approved procedures. Each nuclear facility is required to follow the QA requirements in Appendix B to 10 CFR Part 50. Article 13 describes the QA Program. Criterion V in Appendix B to 10 CFR Part 50, requires that licensees establish measures to ensure that activities that affect quality will be prescribed by appropriate documented instructions, procedures, or drawings. Revision 3 to NRC Regulatory Guide 1.33 provides supplemental guidance. The rule that addresses the need to perform maintenance according to approved procedures is 10 CFR 50.65, and 10 CFR 50.65(a)(4) requires licensees to assess and manage the increase in risk that may result from proposed maintenance activities.

19.4 Procedures for Responding to Anticipated Operational Occurrences and Accidents

The documents providing recommendations and guidance on procedures for responding to anticipated operational occurrences and accidents are NUREG-0737, "Clarification of TMI Action Plan Requirements," issued November 1980; NUREG-0737, Supplement 1, "Requirements for Emergency Response Capability," issued January 1983; and NUREG-0899, "Guidelines for the Preparation of Emergency Operating Procedures," issued August 1982.

After the 1979 accident at TMI Unit 2, the NRC issued orders requiring licensees to develop procedures for coping with certain plant transients and postulated accidents. It also issued NUREG-0737 in 1980 and Supplement 1 to that document in 1983, which recommend that licensees develop procedures to cope with accidents and transients that are caused by initiating events analyzed in the final safety analysis report with multiple failures of equipment.

NUREG-0899 gives programmatic guidance for developing emergency operating procedures. To ensure that proper procedures had been developed to respond to plant transients and accidents, the NRC reviewed each plant using the guidance in NUREG-0800, Section 13.5.2.

19.5 Availability of Engineering and Technical Support

The NRC's Reactor Oversight Process, discussed in Article 6, includes techniques to ensure that adequate engineering and technical support is available throughout the lifetime of a nuclear installation. Several of the inspection procedures focus on ensuring the maintenance of adequate support programs. Licensees also report performance indicators. Depending on inspection findings and performance indicators, the NRC conducts additional inspections to focus on the causes of the performance problems as prescribed by the Reactor Oversight Process Action Matrix.

19.6 Incident Reporting

Two of the many elements contributing to the safety of nuclear power are emergency response and the feedback of operating experience into plant operations. The licensee event reporting requirements of 10 CFR 50.72, "Immediate Notification Requirements for Operating Nuclear Power Reactors," and 10 CFR 50.73, "Licensee Event Report System," help to achieve these, as 10 CFR 50.72 provides for immediate notification requirements via the emergency notification system, and 10 CFR 50.73 provides for 60-day written licensee event reports.

The NRC staff uses the information reported under these regulations in responding to emergencies, monitoring ongoing events, confirming licensing bases, studying potentially generic safety problems, assessing trends and patterns of operational experience, monitoring performance, identifying precursors of more significant events, and providing operational experience to the industry.

The NRC modified these rules in 1992 and 2000 to delete reporting requirements for some events that were determined to be of little or no safety significance. The modified rules continue to provide the Commission with reports of significant events for which the NRC may need to act to maintain or improve reactor safety or to respond to heightened public concern. The modified rules also better align requirements on event reporting with the type of information that the NRC needs to carry out its safety mission. The NRC issued Revision 2 to NUREG-1022, "Event Reporting Guidelines, 10 CFR 50.72 and 50.73," in October 2000, concurrently with the rule changes.

NUREG-1022 is structured to assist licensees in prompt and complete reporting of specified events and conditions. It specifically discusses general issues that have been difficult to implement in the past such as engineering judgment, time limits for reporting, multiple failures and related events, deficiencies discovered during licensee engineering reviews, and human performance issues. It also includes a comprehensive discussion of each specific reporting criterion with illustrative examples and definitions of key terms and phrases.

Event reporting under these rules since 1984 has contributed significantly to focusing the attention of the NRC and the nuclear industry on the lessons learned from operating experience to improve reactor safety. Over the years, decreasing trends in the number of reactor transients and significant events and improvements in reactor safety system performance have been evident.

19.7 Programs to Collect and Analyze Operating Experience

The NRC revised its Operating Experience Program in 2005, as described in the introduction to this report. Upon launching the revised Operating Experience Program, the NRC implemented a number of recommendations concerning better defined roles and responsibilities, a central clearinghouse, and improved collection, storage, and retrieval of information on operating experience.

The Operating Experience Program has four phases, which address all attributes of an effective operating experience program. Management Directive 8.7, "Reactor Operating Experience

Program," (September 28, 2006) explains these phases in detail. This directive also delineates the roles and responsibilities for all participants in the Operating Experience Program and explains the need to periodically assess the program effectiveness. The definition of each phase and the significant program activities and changes under each phase are as follows:

- Phase 1—The first phase of the operating experience process involves collecting, storing, and making operating experience information available to the NRC staff. Through information technology, the NRC has made significant advances in this area, enabling staff to locate and evaluate operating experience information with ease. The collected operating experience includes those inputs considered new information about recent events or conditions at a plant, as well as previously "analyzed" information. Licensees responding to regulatory reporting requirements provide most of the new information. Other sources include NRC inspection reports, INES events, the Incident Reporting System, and other internally generated reports on operating experience. The previously analyzed information contains insights and lessons learned related to the subject operating experience topic. Sources of this type of information include generic communications, inspection findings, INPO reports, and other studies and reports related to operating experience.

- Phase 2—In this phase, the clearinghouse screens a new piece of operating experience information to determine if it has potential significance. The NRC has formalized the screening process through the program guidance documents to ensure a systematic approach to reviewing operating experience. The staff applies a set of screening guidelines that considers risk and qualitative factors, such as potential generic implications, adverse trends, or new phenomena, to screen in those operating experience inputs that are potentially significant and deserving of a more detailed evaluation. Operating experience information screened in for further evaluation becomes a formal assignment, and a clearinghouse staff member gathers additional information to prepare to evaluate the issue. The staff screens out operating experience information that does not meet any of the screening guidelines but may communicate this information to cognizant technical experts or inspection staff. The staff also tracks such information to identify any adverse trends.

- Phase 3—After operating experience information is screened in and communicated to various stakeholders, clearinghouse staff or other technical staff evaluate it to clearly determine its significance for plant operation and safety. The purpose of the evaluation is to glean insights and lessons learned that can be applied toward agency action. The evaluation determines the risk significance and/or identifies other safety or agency concerns associated with the information. The staff generates a report documenting any insights gained and recommending appropriate ways to apply the lessons learned to future regulatory activities. These evaluations have supported improved communication and integration between the clearinghouse, the technical staff, and the regional offices.

- Phase 4—Once the assigned staff member evaluates the screened-in item and recommends further action, the clearinghouse management decides, in consultation with other appropriate NRC managers when necessary, whether to adopt the recommendations. Identified options for applying the lessons learned consist of (1) communicating operating experience lessons learned to various internal and/or external stakeholders through reports, briefings, email listservs, or generic communications, (2)

138

taking regulatory action through a generic communication to require responses from the licensees or issuing orders for actions, and (3) influencing agency programs such as inspection, oversight, licensing, incident response, security, rulemaking, and research. Application always involves communication of the issue to internal stakeholders. Less common outcomes of recommendations are rulemaking or transfer to the agency generic safety issues program.

19.8 Radioactive Waste

The NRC has regulations and guidance for nuclear power reactor licensees to help ensure the safe management and disposal of low-level radioactive waste. Onsite low-level waste must be managed in accordance with the NRC regulations in 10 CFR Part 20 and 10 CFR Part 50. For instance, Subpart K, "Waste Disposal," of 10 CFR Part 20 deals with the treatment and disposition of radioactive waste as an aspect of licensee operations. In addition, GL 81-38, "Storage of Low-Level Radioactive Wastes at Power Reactor Sites," dated November 10, 1981, provides guidance on measures for ensuring the safe storage of low-level waste.

Notwithstanding the preceding regulations and guidance, the economics of waste disposal in the United States have encouraged practices to minimize radioactive waste. In the past decade or so, disposal costs have risen significantly, and volumes of waste produced have decreased greatly as operations technology evolves. Nuclear power reactors now generate only small amounts (about 1000–2000 cubic feet per unit) of operational waste each year.

For storage, waste is put into a form that is stable and safe to minimize the likelihood that it will migrate (e.g., if it were a liquid). Waste that is put into storage is in a form that is suitable for disposal, or at least a form that can be made suitable for disposal. The NRC has specific regulations for the storage of greater than Class C low-level waste produced by nuclear power reactors in 10 CFR Part 72. For designing and operating low-level waste disposal facilities, the NRC has detailed regulations in 10 CFR Part 61.

The U.S. Government addresses the spent fuel and radioactive waste programs, including high-level waste, in detail in a report prepared to satisfy the reporting requirements of the Joint Convention on the Safety of Spent Fuel Management and on the Safety of Radioactive Waste Management. The latest report (DOE/EM-0654, Revision 1, October 2005) is available on the DOE Environmental Management Web site.

APPENDIX A NRC STRATEGIC PLAN 2004–2009

NRC Major Challenges for the Future

The U.S. Nuclear Regulatory Commission (NRC) identified major challenges for the future in its strategic plan for 2004–2009; those that apply to the reactor safety arena are listed below. The NRC is currently working on a new strategic plan.

The Changing Regulatory Environment

The many industries that use radioactive materials are changing, particularly with regard to nuclear safety, security and emergency preparedness, risk-informed, performance-based regulations, energy production, and waste management, creating challenges that must be met. The section below describes changes expected within the next 5 years.

- NRC strategic initiatives will significantly emphasize strengthening the interrelationship among safety, security, and emergency preparedness.

- The majority of operating nuclear power plants will have applied for license renewal to help meet the country's demand for energy. A primary challenge is to monitor, manage, and control the effects of aging so that safety is ensured for the renewal period.

- The U.S. Department of Energy (DOE) will apply to construct and operate the country's high-level radioactive waste repository. The timing of this action will challenge the allocation of the NRC's resources.

- The U.S. nuclear power industry will show a growing interest in licensing and constructing new nuclear power plants to meet the Nation's demand for energy. Challenges include analyzing in detail the vulnerability to accidents and security compromises, as well as developing inspections, tests, analyses, and acceptance criteria for construction

- The NRC, Agreement States (described in Article 8), and licensees will continue to devote increasing attention to the security of radioactive materials and facilities. The primary challenge facing the NRC is to emerge from the period of uncertainty in post-September 11 security requirements; determine what long-term security provisions are necessary; and revise regulations, orders, and internal procedures as necessary to ensure public health and safety and the common defense and security in an elevated threat environment.

- The NRC will continue to see increased requirements to coordinate with a wide array of Federal, State, and local agencies related to homeland security and emergency planning. The NRC currently conducts emergency preparedness exercises that involve a wide array of governmental agencies and emergency response personnel and use cooperative intergovernmental relationships to balance and inform national response capabilities.

• The regulatory climate is expected to adjust to both internal and external factors (described below). Challenges include materials degradation at nuclear power plants, new and evolving technologies, and continuing review of ongoing operational experience.

Key External Factors

The NRC's ability to achieve its goals depends on a changing equation of industry operating experience, national priorities, market forces, and availability of resources. The following section discusses significant external factors, none of which the NRC can control but all of which could affect the agency's ability to achieve its strategic goals.

Receipt of New Reactor Operating License Applications. The U.S. nuclear industry has indicated a new and growing interest in licensing and constructing new nuclear power plants. If the NRC receives a substantial increase in new reactor operating license applications beyond those currently anticipated, the agency would have to significantly reallocate resources to review applications in a timely manner and inspect construction activities. In addition, the high level of public interest likely to be associated with such applications would require significant efforts by the NRC to keep stakeholders informed and involved in the licensing process.

Significant Operating Incident (Domestic or International). A significant safety incident could cause an unexpected increase in safety and security requirements that would likely change the agency's focus on initiatives related to its five goals until the situation stabilized. Because NRC stakeholders (including the public) are highly sensitive to many issues regarding the use of radioactive materials, even events of relatively minor safety or security significance can sometimes require a response that consumes considerable agency resources.

Significant Terrorist Incident. A significant terrorist incident anywhere in the United States could significantly alter the Nation's priorities. This, in turn, could affect significance levels, a need for new or changed security requirements, or other policy decisions that might impact the NRC, its partners, and the industry it regulates. In particular, the impact on State regulatory and enforcement authorities might affect their ability to work with the NRC in achieving its goals. A significant terrorist incident at a nuclear facility or activity anywhere in the world would likely cause similar changes in the NRC's priorities and potentially in U.S. policy regarding export activities, the NRC's role in international security, and/or requirements for security at U.S. nuclear power plants.

Timing of the DOE Application and Related Activities for the High-Level Waste Repository at Yucca Mountain. The proposed repository for spent nuclear fuel represents a major effort for the NRC in planning, review, analysis, and ultimate decision making regarding the licensing of the facility. The agency has begun to ramp up this effort to respond to DOE preapplication activities. The timing of the Department's actions will heavily influence the NRC's resource allocation decisions over the next several years. Acceleration or delay in the DOE activities will most likely require reprogramming of NRC resources, which may affect other programs that are directly associated with achieving the agency's goals.

Homeland Security Initiatives. Emergency preparedness activities with Federal, State, and local agencies continue to increase in scope and number. This affects the agencies' priorities and workloads. As more resources are diverted to external coordination activities, previous work activities must be reprioritized.

Legislative Initiatives. Many legislative initiatives under consideration by Congress could have a major impact on the NRC. In particular, pending energy legislation, if enacted, would affect the agency's priorities and workload. Increasing interest in diversified sources of energy and energy independence could cause an increase in license applications for nuclear power plants. Any attendant increase in resources devoted to license review and analysis might affect the agency's ability to achieve its goals for the planning period.

APPENDIX B NRC MAJOR MANAGEMENT CHALLENGES FOR THE FUTURE

By law, the Inspector General of each Federal agency (discussed in Article 8) is to describe what he or she considers to be the most serious management and performance challenges facing the agency and assess the agency's progress in addressing those challenges. Accordingly, the Inspector General of the NRC prepared his annual assessment of the major management challenges confronting the agency. The latest report, published in October 2006, can be found on the NRC's public Web site.

In his assessment, the Inspector General defined serious management challenges as "mission-critical areas or programs that have the potential for a perennial weakness or vulnerability that, without substantial management attention, would seriously impact agency operations or strategic goals." The most serious management challenges facing the NRC may be, but are not necessarily, areas that are problematic for the agency. The challenges identified represent critical areas or difficult tasks that warrant high-level management attention. In the 2006 report, the Inspector General identified the following nine management challenges to be the most serious as of September 30, 2006. They are not ranked in order of importance. Eight of the nine challenges are essentially the same as those highlighted in the previous U.S. National Report. In 2006, the Inspector General identified a new challenge, titled Ability to Meet the Demand for Licensing New Reactors and removed the challenge Intra-Agency Communication.

Challenge 1: Protection of nuclear material used for civilian purposes

This challenge, which concerns materials control and accounting, is outside the scope of this report and is not covered here.

Challenge 2: Protection of information

NRC employees often generate and work on sensitive information that needs to be protected. Such information can be sensitive unclassified information or classified national security information that is contained in written documents and electronic databases. As a result of ongoing terrorist activity worldwide, the NRC continually reexamines its document control policies. The NRC faces the challenge of balancing the need to protect sensitive information from inappropriate disclosure against its goal of openness in the agency's regulatory processes. In 2006, the NRC made various efforts to protect sensitive information, including personal information, from inappropriate disclosure.

Challenge 3: Development and implementation of a risk-informed and performance-based regulatory oversight approach

The NRC faces the challenge of integrating probabilistic risk assessment (PRA) into regulatory decisionmaking. In fiscal year 2006, the NRC initiated an effort to address the quality of PRAs and develop standard regulatory risk-informed activities. However, full implementation of PRA quality standards will take a number of years.

The NRC has made progress in implementing a risk-informed and performance-based approach at the Nation's 104 operating commercial nuclear power reactors. For example, the agency has combined its Reactor Inspection Program and Reactor Performance Assessment Program to implement the revised Reactor Oversight Process. An integral part of the Reactor Oversight Process is the baseline inspection program that was developed using a risk-informed approach to determine a list of areas to inspect within the seven established cornerstones of safety.

Application of the risk-informed, performance-based approach in the baseline inspection program requires continual refinement. Because it is a living program, the agency dedicates resources to continually reassess and modify it as necessary based on operating experience and industry performance. A recent Reactor Oversight Process self-assessment recognized that regional inspection resources warrant a sizeable increase in staff for the next few years. Potential shortfalls in inspection resources pose a challenge to the agency's ability to ensure that the risk-informed, performance-based approach applied in the baseline inspection program is up to date and reflects lessons learned.

Challenge 4: Ability to modify regulatory processes to meet a changing environment

The NRC faces the challenge of maintaining its core regulatory programs while adapting to emerging changes in its regulatory environment. These changes are listed in the NRC's Strategic Plan. One change is of such significance that the Inspector General has isolated it as a separate challenge (see Challenge 9). The anticipated workload associated with gearing up to receive license applications for new reactors will strain the NRC's current resources. Preparing for the anticipated burden on resources intensifies the challenges posed by other changes in the NRC's regulatory environment. In particular, the NRC must be able to adapt to the following:

- uncertainty in the expected number of applications for license renewals submitted by industry in response to the Nation's demand for energy production

- a heightened public focus on license renewals resulting in contentious hearings

- uncertainty in the expected number of licensee requests to increase power levels

Reactor License Renewal. The NRC's license renewal program is one of the major elements of its regulatory work. The NRC could receive approximately 25 to 30 additional applications to renew operating licenses over the next several years. Because the decision to seek a renewal is the responsibility of the nuclear power plant owner(s), anticipating the number of applications presents a challenge to the NRC. Recent agency experience reflects industry's strong interest in license renewal. Additionally, the NRC will encounter challenges related to a heightened public interest in license renewals that may lead to more contentious hearings. Until 2006, it was unlikely for the NRC to grant hearings on license renewals. In 2006, however, the agency granted the first two such hearings, and the license renewal staff anticipates more.

Applications To Increase Power Output. As of May 2007, the NRC approved 113 power uprate increases, and 11 are pending review. Over the next 5 years, the NRC expects 27 additional requests, which may affect the ability of NRC staff to maintain established review schedules. To address the increase in power uprate requests, the agency is continuing to improve the process on the basis of lessons learned from completed reviews. The process improvements include more detailed analysis of specific technical issues and related efficiencies. Some of the technical issues include power uprate testing programs and reactor systems methods. Also, the

NRC has implemented more rigorous acceptance reviews for power uprate applications to improve the efficiency of the process.

Challenge 5: Implementation of information resources

The NRC relies on a wide variety of information systems to fulfill its responsibilities. In recent years, the agency has created large databases of publicly available information, including the NRC Web site and the Agencywide Documents Access and Management System (ADAMS) public reading room. The following paragraphs highlight some of the NRC's efforts to strengthen and support the agency's business needs using information technology strategies.

Information Security and Federal Information Security Management Act Compliance. The NRC received a low grade on Federal computer security for 2005. To ensure that the agency's systems have adequate security controls to protect information resources, the NRC has engaged a contractor to enhance agencywide information systems security.

Microsoft Office Deployment. The NRC is developing a plan to deploy Microsoft Office Professional software suite; Microsoft Office products will become the agency's standard within the coming year replacing Corel WordPerfect as the agency's standard word processing format.

ADAMS. The Office of Information Services is planning to update ADAMS and then replace it in 2010. This change will present a major challenge to the NRC. The initial cost of the system exceeded agency estimates, and the system took longer to become operational than anticipated and initially failed to significantly improve document management. The challenge will be to incorporate the lessons learned from the first ADAMS experience into an effective transition to a new system.

Challenge 6: Administration of all aspects of financial management

The NRC must be a prudent steward of its fiscal resources through sound financial management. Financial management challenges include preparation of financial statements in accordance with applicable requirements, financial systems replacement, sound budget formulation planning, and efficient and effective procurement operations.

Challenge 7: Communication with external stakeholders throughout the NRC's regulatory activities

The NRC's strategic goal to ensure openness expressly recognizes that the public must be informed about, and have a reasonable opportunity to participate in, the regulatory process. The NRC states that public involvement in, and information about, its activities is the cornerstone of strong, fair regulation of the nuclear industry, and therefore, provides opportunities for citizens to be heard.

Owing to the nature of its business, the agency needs to interact with a diverse group of external stakeholders (e.g., Congress, the general public, other Federal agencies, and various industry and citizen groups) with clear, accurate, and timely information about its regulatory activities.

The NRC enhanced its outreach to external stakeholders in several ways. The agency responded to an extraordinarily high number of stakeholder requests for more information and to numerous congressional inquiries. The agency also conducted extensive interviews with the media and meetings with residents of local communities and State and local government officials to discuss new initiatives, reported events, and other significant regulatory activities.

The NRC encourages public participation and comments applicable to new reactor licensing activities through open meetings, Commission meetings, advisory committee meetings, and other opportunities open to the public. In addition, public meetings between NRC's technical staff and applicants or licensees are open to interested members of the public.

In this post-September 11 environment, the NRC continues to face challenges in determining an appropriate balance between its strategic goal of openness and the need to protect sensitive information. The agency has traditionally been committed to the principles of openness, fairness, and due process. In addition, the Freedom of Information Act requires Federal agencies to make information available to the general public by request or through automatic disclosure of certain types of information.

Challenge 8: Managing human capital

The NRC continues to be challenged by growth in new work at a time when senior experts are increasingly eligible to retire. To mitigate the impact of the challenge, the agency established a Human Capital Council to find, attract, and retain staff members who possess critical skills; implemented human capital provisions of the Energy Policy Act; identified staffing/training and development needs; moved forward with knowledge management strategies; and monitored the attrition rate.

Challenge 9: Ability to meet the demand for licensing new reactors

There is a growing list of U.S. licensees that are considering new nuclear power plant construction. These licensees intend to apply for early site permits, combined licenses, and design certifications. Title 10, Part 52, "Early Site Permits; Standard Design Certifications; and Combined Licenses for Nuclear Power Plants," of the *Code of Federal Regulations* (10 CFR Part 52) outlines the NRC's licensing process. The agency is involved in several significant activities to ensure that it is prepared to review the first of the combined operating license (COL) applications which is expected in 2007–2008. These activities include the following:

- reviewing industry's guidelines for a COL application

- determining what actions are necessary to prepare for receipt of a COL application

- assessing rulemaking activities for the licensing process

- reviewing early site permit applications

- developing the Multinational Design Approval Program with international regulators, which will take advantage of worldwide nuclear safety, licensing, and operating experience

The NRC has already certified some new reactor designs under the new 10 CFR Part 52 licensing process. Under this approach, the agency preapproves or certifies new reactor designs and allows licensees to apply for an early site permit and/or a COL using one of the preapproved designs. Also, the NRC intends to apply a design-centered approach to facilitate effective, efficient, and timely review of multiple COL applications. This approach streamlines and shortens the NRC review process.

Although the 10 CFR Part 52 application process has advantages for both the NRC and the nuclear industry, it nevertheless represents a significant challenge because of the increased workload and pressure on the agency to create the infrastructure necessary to support review of new plant licensing applications.

As the NRC enters a new era of reactor regulation, it faces many challenges. In addition to ongoing license renewal activities, the agency will face the first round of new reactor applications since 1978. The NRC estimates that it may receive 20 or more applications in the coming years and that upward of 450 new staff positions will be needed to review these applications.

Coinciding with the dramatic increase in regulatory responsibilities will be the retirement of many senior staff members who have experience in licensing reactors from the 1960s, 1970s, and 1980s. The agency's ability to effectively review and license the new generation of commercial nuclear reactors will depend significantly on how well employees, new to the process, are trained and developed into effective reviewers and regulators at the staff and senior management level. Furthermore, construction oversight of future plants will be equally or even more challenging.

The review of new applications involving new reactor technologies, a new licensing process, and new staff untested in this realm necessitates a strong control process to ensure that the agency meets its review and licensing objectives. Specific challenges include the following:

- Project Management—Effective technical and communications skills are essential for the focal point (the project manager) of NRC and licensee interactions.

- Construction Inspection Oversight—The NRC must reinstitute this program which has been dormant for many years.

- Technical Review Process—The NRC must have a defined process for ensuring that all requisite technical reviews are conducted, documented, and approved.

- Standard Review Plan—As it did for the previous generation of reactors, the NRC must have a comprehensive Standard Review Plan for examining a license application. Additionally, consistent implementation of the Standard Review Plan is vital.

- Safety Evaluation Reports—The agency needs a solid process for compiling its regulatory examination into a safety evaluation report. This report reflects the agency's conclusion about a plant's ability to operate safely. It is essential that such conclusions be documented and approved.

APPENDIX C REFERENCES

American National Standards Institute (ANSI), ANSI N18.7-1976, "Administrative Controls and Quality Assurance for the Operational Phase of Nuclear Power Plants," 1976

American Society of Mechanical Engineers (ASME), "Boiler and Pressure Vessel Code," Section XI

————, ASME-RA-5-2002, April 2002, and Addendum 1, December 2003, and Addendum 2, December 2005

Federal Emergency Management Agency (FEMA), FEMA-REP-1 (See NUREG-0654.)

————, Federal Policy Statement on Potassium Iodide Prophylaxis, January 2002

International Atomic Energy Agency, Safety Series No. 75-INSAG-4, "Safety Culture," Vienna, 1991

————, TECDOC-953, "Method for the Development of Emergency Response Preparedness for Nuclear or Radiological Accidents," Vienna, 1997

————, TECDOC-955, "Generic Assessment Procedures for Determining Protective Actions During a Reactor Accident," Vienna, 1997

International Commission on Radiological Protection, ICRP Publication 26, "Recommendations of the International Commission on Radiological Protection (Adopted January 17, 1977)," Oxford, Pergamon Press, 1991

————, ICRP Publication 30, "Limits of Intakes of Radionuclides by Workers," eight volumes, Oxford, Pergamon Press, 1978–1982

National Council on Radiation Protection and Measurements, NCRP Report No. 91, "Recommendations on Limits for Exposure to Ionizing Radiation," June 1987

Executive Order 12898, "Federal Actions To Address Environmental Justice in Minority and Low-Income Populations," 59 FR 7629, February 1994

Nuclear Energy Institute, NEI 04-04, "Cyber Security Programs for Power Reactors," November 18, 2005.

Homeland Security Presidential Directive 5 (HSPD 5), "Management of Domestic Incidents," March 4, 2003. Available at http://www.whitehouse.gov/XXXXX as of June 2007.

National Energy Policy, March 27, 200X. Available at http://www.whitehouse.gov as of June 2007.

U.S. Congress, Atomic Energy Act of 1954, as amended, 42 U.S.C. 2011 et seq.

————, Debt Collection Improvement Act, 5 U.S.C. 5514 et seq.

| ————, DOE Organization Act, 42 U.S.C. 7101

————, Energy Policy Act of 2005, 16 U.S.C. 797 note et seq.

————, Energy Reorganization Act of 1974, as amended, 42 U.S.C. 5801 et seq.

| ————, Federal Civil Penalties Inflation Adjustment Act, 28 U.S.C. 2461

| ————, Federal Food, Drug, and Cosmetic Act, 21 U.S.C. 301

————, Freedom of Information Act, 5 U.S.C. 552 et seq.

————, Hobbs Act (See Administrative Orders Review Act.)

| ————, Homeland Security Act of 2002, 6 U.S.C. 101

————, National Environmental Policy Act of 1969, as amended, 42 U.S.C. 4321 et seq.

————, Nuclear Non-Proliferation Act of 1978, 22 U.S.C. 3201 et seq.

————, Price-Anderson Act of 1957, 42 U.S.C. 2012 et seq.

————, Uranium Mill Tailings Radiation Control Act of 1978, 42 U.S.C. 6907 et seq.

| **U.S. Department of Energy**, DOE/EM-0654, Revision 1, "Second National Report for the Joint
| Convention on the Safety of Spent Fuel Management and on the Safety of Radioactive Waste
| Management." DOE in cooperation with the U.S. NRC, U.S. Environmental Protection Agency,
| and U.S. Department of State. DOE: Washington, DC. October 2005.

| **U.S. Department of Homeland Security,** National Response Plan, March 18, 2003

U.S. Environmental Protection Agency, EPA-400-R-92-001, "Manual of Protective Action
Guides and Protective Actions for Nuclear Incidents," May 1992

U.S. Nuclear Regulatory Commission, "Agreement Between the Government of the United
States of America and the Government of Canada on Cooperation in Comprehensive Civil
Emergency Planning and Management" and "Administrative Arrangement Between the United
States Nuclear Regulatory Commission and the Atomic Energy Control Board of Canada for
Cooperation and the Exchange of Information in Nuclear Regulatory Matters," June 21, 1989

————, "Agreement for the Exchange of Information and Cooperation in Nuclear Safety Matters"
and "Implementing Procedure for the Exchange of Technical Information and Cooperation in
Nuclear Safety Matters Between the Nuclear Regulatory Commission of the United States of
America and the Comision Nacional de Seguridad Nuclear y Salvaguardias of Mexico,"
October 6, 1989

————, "Availability and Adequacy of Design Bases Information at Nuclear Power Plants; Policy Statement," 57 FR 35455, August 10, 1992

————, Bulletin 2003-02, "Leakage from Reactor Pressure Vessel Lower Head Penetrations and Reactor Coolant Pressure Boundary Integrity," August 21, 2003

————, "Commission Policy Statement on the Systematic Evaluation of Operating Nuclear Power Reactors," 49 FR 45112, November 1984

————, Draft Regulatory Guide 1145, "Combined License Applications for Nuclear Power Plants (LWR Edition)," September 2006

————, "Final Policy Statement on Technical Specification Improvement for Nuclear Power Reactors," 58 FR 39132, July 22, 1993

————, Generic Letter 81-38, "Storage of Low Level Radioactive Wastes at Power Reactor Sites," November 10, 1981

————, Generic Letter 2004-02, "Potential Impact of Debris Blockage on Emergency Recirculation during Design Basis Accidents at Pressurized-Water Reactors," September 13, 2004, ADAMS Accession No. ML042360586

————, Generic Letter 2006-02, "Grid Reliability and the Impact on Plant Risk and the Operability of Offsite Power," February 1, 2006

————, Information Notice 97-78, "Crediting of Operator Actions in Place of Automatic Actions and Modifications of Operator Actions, Including Response Times," October 23, 1997

————, Information Notice 2004-05, "Spent Fuel Pool Leakage to Onsite Groundwater," March 3, 2004

————, Information Notice 2005-26, "Results of Chemical Effects Head Loss Tests in a Simulated PWR Sump Pool Environment," September 16, 2005, ADAMS Accession No. ML052570220

————, Information Notice 2005-26, Supplement 1, "Additional Results of Chemical Effects Tests in a Simulated PWR Sump Pool Environment," January 20, 2006, ADAMS Accession No. ML060170102

————, Information Notice 2006-13, "Ground-Water Contamination Due to Undetected Leakage of Radioactive Water," July 10, 2006

————, Information Notice 2006-30, "Summary of Fitness-for-Duty Program Performance Reports for Calendar Years 2004 and 2005," December 21, 2006

————, Inspection Manual Chapter 0350, "Oversight of Operating Reactor Facilities in a Shutdown Condition with Performance Problems"

————, Inspection Manual Chapter 2501, "Early Site Permits"

————, Inspection Manual Chapter 2502, "Pre-Combined License Phase"

————, Inspection Manual Chapter 2503, "ITAAC"

————, Inspection Manual Chapter 2504, "Non-ITAAC Inspections"

————, Inspection Manual Chapter 2509, "Browns Ferry Unit 1 Restart Project Inspection Program"

————, Inspection Procedure 41500, "Training and Qualification Effectiveness"

————, Inspection Procedure 42001, "Emergency Operating Procedures"

————, Inspection Procedure 42700, "Plant Procedures"

————, Inspection Procedure 71111.11, "Licensed Operator Requalification Program"

————, Inspection Procedure 71152, "Identification and Resolution of Problems"

————, Inspection Procedure 95003, "Supplemental Inspection of Repetitive Degraded Cornerstones, Multiple Degraded Cornerstones, Multiple Yellow Inputs, or One Red Input"

————, Management Directive 8.7, "Reactor Operating Experience Program," September 28, 2006

————, Management Directive 10.159, "The NRC Differing Professional Opinions Program," May 16, 2004

————, NUREG-0396, "Planning Basis for the Development of State and Local Government Radiological Emergency Response Plans in Support of Light Water Nuclear Power Plants, EPA-520/1-78/016," December 1978

————, NUREG-0654, "Criteria for Preparation and Evaluation of Radiological Emergency Response Plans and Preparedness in Support of Nuclear Power Plants," FEMA-REP-1, Rev. 1, November 1980

————, NUREG-0654, Supplement 3, "Criteria for Protective Action Recommendations for Severe Accidents (Draft Report for Interim Use and Comment)," Rev. 1, July 1996

————, NUREG-0700, "Human System Interface Design Review Guideline," Rev. 2, May 2002

————, NUREG-0711, "Human Factors Engineering Program Review Model," July 1994

————, NUREG-0713, "Occupational Radiation Exposure at Commercial Nuclear Power Reactors and Other Facilities," Vol. 24, October 2003

————, NUREG-0728, "NRC Incident Response Plan," Rev. 4, November 23, 2003

————, NUREG-0737, "Clarification of TMI Action Plan Requirements," November 1980

————, NUREG-0737, Supplement 1, "Requirements for Emergency Response Capability," January 1983

————, NUREG-0800, "Standard Review Plan for the Review of Safety Analysis Reports for Nuclear Power Plants," 1981, 1984, and 1987 (formerly NUREG-75/087)

————, NUREG-0899, "Guidelines for the Preparation of Emergency Operating Procedures," August 1982

————, NUREG-0933, "A Prioritization of Generic Issues," October 2006

————, NUREG-1021, "Operator Licensing Examination Standards for Power Reactors," Rev. 8, April 1999

————, NUREG-1022, "Event Reporting Guidelines, 10 CFR 50.72 and 50.73," Rev. 2, October 2000

————, NUREG-1150, "Severe Accident Risks: An Assessment for Five U.S. Nuclear Power Plants," December 1990

————, NUREG-1220, "Training Review Criteria and Procedures," Rev. 1, January 1993

————, NUREG-1437, "Generic Environmental Impact Statement for License Renewal of Nuclear Power Plants," May 1996

————, NUREG-1465, "Accident Source Terms for Light-Water Nuclear Power Plants," February 1995

————, NUREG-1555, "Standard Review Plans for Environmental Reviews for Nuclear Power Plants, Supplement 1: Operating License Renewal," March 2000

————, NUREG-1577, Revision 1, "Standard Review Plan on Power Reactor Licensee Financial Qualifications and Decommissioning Funding Assurance," February 1999

————, NUREG-1600, "General Statement of Policy and Procedures for NRC Enforcement Actions (Enforcement Policy)," July 2000

————, NUREG-1650, "U.S. National Report for the Convention on Nuclear Safety," September 2001, September 2004

————, NUREG-1764, "Guidance for the Review of Changes to Human Actions, Draft Report for Comment," December 2002

————, NUREG-1791, "Guidance for Assessing Exemption Requests from the Nuclear Power Plant Licensed Operator Staffing Requirements Specified in 10 CFR 50.54(m)," July 2005

————, NUREG-1800, "Standard Review Plan for Review of License Renewal Applications for Nuclear Power Plants," July 2001

————, NUREG-1801, "Generic Aging Lessons Learned (GALL) Report," July 2001

————, NUREG/CR-2850, "Dose Commitments Due to Radioactive Releases from Nuclear Power Plant Sites in 1992," Vol. 14, March 1996

————, NUREG/CR-6838, "Technical Basis for Assessing Exemptions from Nuclear Power Plant Licensed Operator Staffing Requirements 10 CFR 50.54(m)," February 2004

————, NUREG/CR-6847 "Cyber security Self-Assessment Method for U.S. Nuclear Power Plants," October 2004

————, "OIG 2005 Survey of NRC's Safety Culture and Climate," OIG-06-A-08, February 10, 2006

————, "Policy Statement on Deferred Plants," (52 FR 38077, October 14, 1987)

————, "Policy Statement on Use of PRA Methods in Nuclear Activities," 60 FR 42623, August 16, 1995

————, RG 1.3, "Assumptions Used for Evaluating the Potential Radiological Consequences of a Loss-of-Coolant Accident for Boiling-Water Reactors," Rev. 2, June 1974

————, RG 1.4, "Assumptions Used for Evaluating the Potential Radiological Consequences of a Loss-of-Coolant Accident for Pressurized-Water Reactors," Rev. 2, June 1974

————, RG 1.8, "Personnel Selection and Training," March 1971 and RG 1.8, "Qualification and Training of Personnel for Nuclear Power Plants," Rev. 2, April 1987

————, RG 1.20, "Comprehensive Vibration Assessment Program for Reactor Internals During Preoperational and Initial Startup Testing," November 2006

————, RG 1.33, "Quality Assurance Program Requirements (Operations)," Rev. 3, November 1980

————, RG 1.70, "Standard Format and Content of Safety Analysis Reports for Nuclear Power Plants," Rev. 3, November 1978

————, RG 1.84, "Design, Fabrication, and Materials Code Case Acceptability, ASME Section III," March 2003

————, RG 1.101, "Emergency Planning and Preparedness for Nuclear Power Plants," Rev. 4, July 2003

————, RG 1.145, "Atmospheric Dispersion Models for Potential Accident Consequence

Assessments at Nuclear Power Plants," Rev. 1, November 1982 (reissued with corrected page 1.145-7, February 1983)

———, RG 1.147, "Inservice Inspection Code Case Acceptability, ASME Section XI, Division 1," January 2004

———, RG 1.165, "Identification and Characterization of Seismic Sources and Determination of Safe Shutdown Earthquake Ground Motion," March 1997

———, RG 1.174, "An Approach for Using Probabilistic Risk Assessment in Risk-Informed Decisions on Plant-Specific Changes to the Licensing Basis,"
July 1998 and Rev. 1, November 2002

———, RG 1.177, "An Approach for Plant-Specific, Risk-Informed Decisionmaking: Technical Specifications," August 1998

———, RG 1.178, "An Approach for Plant-Specific Risk-Informed Decisionmaking for Inservice Inspection of Piping," September 2003

———, RG 1.183, "Alternative Radiological Source Terms for Evaluating Design Basis Accidents at Nuclear Power Reactors," July 2000

———, RG 1.188, "Standard Format and Content of Applications to Renew Nuclear Power Plant Operating Licenses," July 2001

———, RG 1.192, "Operation and Maintenance Code Case Acceptability, ASME OM Code," June 2003

———, RG 1.200, "An Approach for Determining the Technical Adequacy of Probabilistic Risk Assessment Results for Risk-Informed Activities," Rev. 1, January 2007

———, RG 1.206, "Combined Operating Licenses for Nuclear Power Plants," June 2007

———, RG 1.208, "A Performance-Based Approach to Define the Site-Specific Earthquake Ground Motion, March 2007

———, RG 4.2, "Preparation of Environmental Reports for Nuclear Power Stations," Rev. 2, July 1976 and Supplement 1, "Preparation of Supplemental Environmental Reports for Applications to Renew Nuclear Power Plant Operating Licenses," August 1991

———, RG 8.8, "Information Relevant to Ensuring That Occupational Radiation Exposures at Nuclear Power Stations Will Be As Low As Is Reasonably Achievable," Rev. 3, June 1978

———, Review Standard (RS-001) for Extended Power Uprates, December 2003

———, Review Standard (RS-002) for Processing Applications for Early Site Permits, December 2003

————, "Safety Goals for the Operation of Nuclear Power Plants: Policy Statement; Republication 51 FR 30028," August 21, 1986

————, SECY-93-087, "Policy, Technical, and Licensing Issues Pertaining to Evolutionary and Advanced Light-Water Reactor Designs," April 2, 1993

————, SECY-06-0164, "NRC Knowledge Management Program," July 25, 2006

————, SECY-06-0208, "Status of the Accident Sequence Precursor Program and the Development of Standardized Plant Analysis Risk Models," October 5, 2006

————, SECY-06-0244, "Final Rulemaking—10 CFR Part 26—Fitness-for-Duty Programs," December 22, 2006

————, SECY-07-0063, "FY 2006 Results of the Industry Trends Program for Operating Power Reactors and Status of Ongoing Development," April 3, 2007

————, "Severe Reactor Accidents Regarding Future Designs and Existing Plants; Policy Statement," 50 FR 32138, August 8, 1985

APPENDIX D ABBREVIATIONS

ADAMS   Agencywide Documents Access and Management System (NRC)
ALARA   as low as reasonably achievable
ANS   American Nuclear Society
ANSI   American National Standards Institute
ASME   American Society of Mechanical Engineers

BPV   boiler and pressure vessel
BWR   boiling-water reactor
CCDP   conditional core damage probability
CEUS   Central and Eastern United States
CFR   *Code of Federal Regulations*
CRGR   Committee To Review Generic Requirements (NRC)
CY   calendar year

DBT   design-basis threat
DHS   U.S. Department of Homeland Security
DOE   U.S. Department of Energy

EPA   U.S. Environmental Protection Agency
EPRI   Electric Power Research Institute
EPU   extended power uprate
ERDA   U.S. Energy Research and Development Administration
ESBWR   Economic and Simplified Boiling-Water Reactor

FDA   U.S. Food and Drug Administration
FEMA   U.S. Federal Emergency Management Agency
FPL   Florida Power and Light
FTE   full-time equivalent
FY   fiscal year

GL   generic letter

I&C   Instrumentation and control
IAEA   International Atomic Energy Agency
ICRP   International Commission on Radiological Protection
IG   Inspector General
IN   information notice
INES   International Nuclear Event Scale
INPO   Institute of Nuclear Power Operations
IP   inspection procedure
IPA   integrated plant assessment
IPE   individual plant examination
IRRS   Integrated Regulatory Review Service
IRRT   International Regulatory Review Team
ISAP   Integrated Safety Assessment Program
ISO   International Organization for Standardization

ITAAC	inspections, tests, analyses, and acceptance criteria
IT/IM	information technology/information management
KM	knowledge management
LAR	license application request
LOCA	loss-of-coolant accident
Mwth	Megawatts Thermal
NCRP	National Council on Radiation Protection and Measurements
NEA	Nuclear Energy Agency
NNAB	National Nuclear Accrediting Board
NRC	U.S. Nuclear Regulatory Commission
NRR	Office of Nuclear Reactor Regulation (NRC)
OM	operation and maintenance
OSART	Operational Safety Assessment Review Team
PRA	probabilistic risk assessment
PS&G	Public Service Electric and Gas Company
PSR	periodic safety review
PWR	pressurized-water reactor
PWSCC	primary water stress-corrosion cracking
QA	quality assurance
RPP	risk-informed and performance-based plan
RS	review standard
SAT	systems approach to training
SE	safety evaluation
SEP	systematic evaluation program
SSC	structure, system, and component
STP	South Texas Project
Sv	sievert
TMI	Three Mile Island
TOPOFF	top officials (emergency response exercise)
TSO	technical support organization
TVA	Tennessee Valley Authority
TXU	Texas Utilities
WENRA	Western European Nuclear Regulators' Association
WBN	Watts Bar Nuclear Plant

APPENDIX E ACKNOWLEDGMENTS

Contributors to this report include the following technical and regulatory experts from the NRC. The project manager was Jon Hopkins; the preparer of the report was Merrilee Banic.

Abu-Eid, Rateb
Alexion, Thomas
Banic, Merrilee
Bavol, Bruce
Brown, Eva
Casto, Gregory
Crockett, Steven
Decker, David
Dinitz, Ira
Dusaniwskyj, Michael
Erlanger, Craig
Fields, Leslie
Ford, William
Frumkin, Daniel
Foster, Jack
Gareri, Mario
Ghasemian, Shahram
Harrison, Donald
Hart, Michelle
Hiltz, Thomas
Hoffman, Stephen
Hopkins, Jon
Huyuk, Douglas
Imboden, Stacy
Jackson, Christopher
James, Lois
Jolicoeur, John
Khanna, Meena
Lauron, Carolyn
Lewis, Robert
Li, Yong
Liu, Christiana
Magruder, Stewart
McConnell, Matthew
Mendiola, Anthony
Milligan, Patricia
Miranda, Samuel
Monninger, John
Morris, Scott
Munson, Cliff
Murphy, Martin
Paige, Jason
Pascarelli, Robert

Perez, Donna-Marie
Prescott, Paul
Ramey-Smith, Ann
Rand, Jennifer
Schwartzman, Jennifer
Scott, Michael
Stuchell, Sheldon
Virgilio, Rosetta
Westreich, Barry
White, Bernard
Williams, Donna
Yoder, Matthew
Zalcman, Barry
Zimmerman, Jacob

161

ANNEX 1 U.S. COMMERCIAL NUCLEAR POWER REACTORS

SOURCE: U.S. Nuclear Regulatory Commission (NRC) and licensee data as compiled by the NRC

RELEVANT ARTICLE: Article 6

Plant Name and Operating Utility	Reactor Design Type	Licensed Power (MWth)	Operating Lifetime
Arkansas Nuclear 1 Entergy Operations	PWR	2568	12/74 05/34
Arkansas Nuclear 2 Entergy Nuclear	PWR	3026	03/80 07/38
Beaver Valley 1 First Energy Nuclear Operating Company	PWR	2689	10/76 01/16
Beaver Valley 2 First Energy Nuclear Operating Company	PWR	2689	11/87 05/27
Braidwood 1 Exelon	PWR	3586.6	07/88 10/26
Braidwood 2 Exelon	PWR	3586.6	10/88 12/27
Browns Ferry 1 Tennessee Valley Authority	BWR	3293	08/74 12/13
Browns Ferry 2 Tennessee Valley Authority	BWR	3458	03/75 06/14
Browns Ferry 3 Tennessee Valley Authority	BWR	3458	03/77 0716
Brunswick 1 Carolina Power and Light, Co.	BWR	2923	03/77 09/16
Brunswick 2 Carolina Power and Light, Co.	BWR	2923	11/75 12/14

Plant Name and Operating Utility	Reactor Design Type	Licensed Power (MWth)	Operating Lifetime
Byron 1 Exelon	PWR	3585.6	09/85 10/24
Byron 2 Exelon	PWR	3586.6	08/87 11/26
Callaway AmerenUE	PWR	3565	12/84 10/24
Calvert Cliffs 1 Nuclear Power Plant, Inc.	PWR	2700	05/75 07/34
Calvert Cliffs 2 Nuclear Power Plant, Inc.	PWR	2700	04/77 08/36
Catawba 1 Duke Energy Power Company, LLC	PWR	3411	06/85 12/43
Catawba 2 Duke Energy Power Company, LLC	PWR	3411	08/86 12/43
Clinton AmerGen Energy Co.	BWR	3473	11/87 09/26
Columbia Generating Station Energy Northwest	BWR	3486	12/84 12/23
Comanche Peak 1 TXU Generation Company LP	PWR	3458	08/90 02/30
Comanche Peak 2 TXU Electric & Gas	PWR	3458	08/93 02/33
Cooper Nebraska Public Power District	BWR	2381	07/74 01/14

Plant Name and Operating Utility	Reactor Design Type	Licensed Power (MWth)	Operating Lifetime
Crystal River 3 Florida Power Corp.	PWR	2568	03/77 12/16
Davis-Besse First Energy Nuclear Operating Company	PWR	2772	07/78 04/17
D.C. Cook 1 Indiana/Michigan Power Co.	PWR	3304	08/75 10/34
D.C. Cook 2 Indiana/Michigan Power Co.	PWR	3468	07/78 12/37
Diablo Canyon 1 Pacific Gas & Electric Co.	PWR	3338	05/85 09/21
Diablo Canyon 2 Pacific Gas & Electric Co.	PWR	3411	03/86 04/25
Dresden 2 Exelon	BWR	2957	06/70 12/29
Dresden 3 Exelon	BWR	2957	11/71 01/31
Duane Arnold Nuclear Management Co.	BWR	1912	02/975 02/14
Edwin I. Hatch 1 Southern Nuclear Operating Co.	BWR	2804	12/75 08/34
Edwin I. Hatch 2 Southern Nuclear Operating Co.	BWR	2804	09/79 06/38
Fermi 2	BWR	3430	01/88

Plant Name and Operating Utility	Reactor Design Type	Licensed Power (MWth)	Operating Lifetime
Detroit Edison Co.			03/25
Ginna Nuclear Power Plant, LLC	PWR	1520	07/70 09/29
Grand Gulf 1 Entergy Operations, Inc.	BWR	3833	07/85 11/24
H.B. Robinson 2 Carolina Power and Light Co.	PWR	2339	03/71 07/30
Hope Creek 1 PSEG Nuclear, LLC	BWR	3339	12/86 04/26
Indian Point 2 Entergy Nuclear Operations	PWR	3216	08/74 09/13
Indian Point 3 Entergy Nuclear Operations	PWR	3216	08/76 12/15
James A. FitzPatrick Entergy Nuclear Operations	BWR	2536	07/75 10/14
Joseph M. Farley 1 Southern Nuclear Operating Co.	PWR	2775	12/77 06/37
Joseph M. Farley 2 Southern Nuclear Operating Co.	PWR	2775	07/81 03/41
Kewaunee	PWR	1772	06/74

Plant Name and Operating Utility	Reactor Design Type	Licensed Power (MWth)	Operating Lifetime
Dominion Energy			12/13
La Salle 1 Exelon	BWR	3489	01/84 04/22
La Salle 2 Exelon	BWR	3489	10/84 12/23
Limerick 1 Exelon	BWR	3458	02/86 10/24
Limerick 2 Exelon	BWR	3458	01/90 06/29
McGuire 1 Duke Energy Power Company, LLC	PWR	3411	12/81 06/41
McGuire 2 Duke Energy Power Company, LLC	PWR	3411	03/84 03/43
Millstone 2 Dominion Generation	PWR	2700	12/75 07/35
Millstone 3 Dominion Generation	PWR	3411	04/86 11/45
Monticello Nuclear Management Co.	BWR	1775	06/71 09/10

Plant Name and Operating Utility	Reactor Design Type	Licensed Power (MWth)	Operating Lifetime
Nine Mile Point 1 Constellation Nuclear	BWR	1850	12/69 08/09
Nine Mile Point 2 Nuclear Station, LLC	BWR	3467	03/88 10/26
North Anna 1 Dominion Generation Operating Utility	PWR	2893	06/78 04/38
North Anna 2 Dominion Generation	PWR	2893	12/80 08/40
Oconee 1 Duke Energy Power Company, LLC	PWR	2568	07/73 02/33
Oconee 2 Duke Energy Power Company, LLC	PWR	2568	09/74 10/33
Oconee 3 Duke Energy Power Company, LLC	PWR	2568	12/74 12/34
Oyster Creek AmerGen Energy Co., LLC	BWR	1930	12/69 04/09
Palisades Nuclear Management Co.	PWR	2565	12/71 03/11
Palo Verde 1 Arizona Public Service Co.	PWR	3990	01/86 12/24
Palo Verde 2 Arizona Public Service Co.	PWR	3990	09/86 12/25
Palo Verde 3 Arizona Public Service Co.	PWR	3990	01/88 03/27

Plant Name and Operating Utility	Reactor Design Type	Licensed Power (MWth)	Operating Lifetime
Peach Bottom 2 Exelon	BWR	3514	07/74 08/33
Peach Bottom 3 Exelon	BWR	3514	12/74 07/34
Perry 1 First Energy Nuclear Operating Company	BWR	3758	11/87 03/26
Pilgrim 1 Entergy Nuclear	BWR	2028	12/72 06/12
Point Beach 1 Nuclear Management Co.	PWR	1540	12/70 10/30
Point Beach 2 Nuclear Management Co.	PWR	1540	10/72 03/33
Prairie Island 1 Nuclear Management Co.	PWR	1650	12/73 08/13
Prairie Island 2 Nuclear Management Co.	PWR	1650	12/74 10/14
Quad Cities 1 Exelon	BWR	2957	02/73 12/32
Quad Cities 2 Exelon	BWR	2957	03/73 12/32
River Bend 1 Entergy Nuclear Operations, Inc.	BWR	3091	06/86 08/25

171

Plant Name and Operating Utility	Reactor Design Type	Licensed Power (MWth)	Operating Lifetime
Salem 1 PSEG Nuclear, LLC	PWR	3459	06/77 08/16
Salem 2 PSEG Nuclear, LLC	PWR	3459	10/81 04/20
San Onofre 2 Southern California Edison Co.	PWR	3438	08/83 02/22
San Onofre 3 Southern California Edison Co.	PWR	3438	04/84 11/22
Seabrook 1 FPL Energy Seabrook	PWR	3587	08/90 10/26
Sequoyah 1 Tennessee Valley Authority	PWR	3411	07/81 09/20
Sequoyah 2 Tennessee Valley Authority	PWR	3411	06/82 09/21
Shearon Harris 1 Carolina Power and Light Co.	PWR	2900	05/87 10/26
South Texas Project 1 STP Nuclear Operating Co.	PWR	3853	08/88 08/27
South Texas Project 2 STP Nuclear Operating Co.	PWR	3853	06/89 12/28
St. Lucie 1 Florida Power & Light Co.	PWR	2700	12/76 03/36
St. Lucie 2 Florida Power & Light Co.	PWR	2700	08/83 04/43

Plant Name and Operating Utility	Reactor Design Type	Licensed Power (MWth)	Operating Lifetime
Summer South Carolina Electric & Gas Co.	PWR	2900	01/84 08/22
Surry 1 Dominion Generation	PWR	2546	12/72 05/32
Surry 2 Dominion Generation	PWR	2546	05/73 01/33
Susquehanna 1 PPL Susquehanna, LLC	BWR	3489	06/83 07/22
Susquehanna 2 PPL Susquehanna, LLC	BWR	3489	02/85 03/24
Three Mile Island 1 AmerGen Energy Co.	PWR	2568	09/74 04/14
Turkey Point 3 Florida Power & Light Co.	PWR	2300	12/72 07/32
Turkey Point 4 Florida Power & Light Co.	PWR	2300	09/73 04/33
Vermont Yankee Entergy Nuclear Northeast	BWR	1912	11/72 03/12
Vogtle 1 Southern Nuclear Operating Co.	PWR	3565	06/87 01/27
Vogtle 2 Southern Nuclear Operating Co.	PWR	3565	05/89 02/29
Waterford 3 Entergy Nuclear Operations, Inc	PWR	3716	09/85 12/24

Plant Name and Operating Utility	Reactor Design Type	Licensed Power (MWth)	Operating Lifetime
Watts Bar 1 Tennessee Valley Authority	PWR	3459	05/96 11/35
Wolf Creek 1 Wolf Creek Nuclear Operating Corp.	PWR	3565	09/85 03/25

PART 3